CAMBRIDGE LIBRARY COLLECTION

Books of enduring scholarly value

Darwin

Two hundred years after his birth and 150 years after the publication of 'On the Origin of Species', Charles Darwin and his theories are still the focus of worldwide attention. This series offers not only works by Darwin, but also the writings of his mentors in Cambridge and elsewhere, and a survey of the impassioned scientific, philosophical and theological debates sparked by his 'dangerous idea'.

The Principles of Descriptive and Physiological Botany

Henslow's importance as Darwin's mentor is well established. He recommended Darwin for the post of naturalist on the Beagle and also encouraged him to read Lyell's pivotal geology text (also reissued in this series). While professor of botany at Cambridge, Henslow nurtured independent inquiry and acute observation in his students. These attributes are evident in this liberally illustrated 1835 book, which also reveals the influence of Candolle's Théorie Elémentaire de la Botanique (1813) and Physiologie Végétale (1832). Henslow's book, like his meticulous research papers and his innovative lectures, included focussed investigations on the nature and stability of 'species'. Charles Darwin paid such close attention that he became known as 'the man who walks with Henslow', and Henslow's teachings were to echo through Darwin's writings, from his jottings in notebooks on the Beagle onward. This reissue gives modern readers easy access to the work of this inspirational scientist.

Cambridge University Press has long been a pioneer in the reissuing of out-of-print titles from its own backlist, producing digital reprints of books that are still sought after by scholars and students but could not be reprinted economically using traditional technology. The Cambridge Library Collection extends this activity to a wider range of books which are still of importance to researchers and professionals, either for the source material they contain, or as landmarks in the history of their academic discipline.

Drawing from the world-renowned collections in the Cambridge University Library, and guided by the advice of experts in each subject area, Cambridge University Press is using state-of-the-art scanning machines in its own Printing House to capture the content of each book selected for inclusion. The files are processed to give a consistently clear, crisp image, and the books finished to the high quality standard for which the Press is recognised around the world. The latest print-on-demand technology ensures that the books will remain available indefinitely, and that orders for single or multiple copies can quickly be supplied.

The Cambridge Library Collection will bring back to life books of enduring scholarly value (including out-of-copyright works originally issued by other publishers) across a wide range of disciplines in the humanities and social sciences and in science and technology.

The Principles of Descriptive and Physiological Botany

John Stevens Henslow

CAMBRIDGE UNIVERSITY PRESS

Cambridge, New York, Melbourne, Madrid, Cape Town, Singapore,
São Paolo, Delhi, Dubai, Tokyo

Published in the United States of America by Cambridge University Press, New York

www.cambridge.org
Information on this title: www.cambridge.org/9781108001861

This edition first published 1835
This digitally printed version 2009

ISBN 978-1-108-00186-1 Paperback

THE

CABINET CYCLOPÆDIA.

CONDUCTED BY THE

REV. DIONYSIUS LARDNER, LL.D. F.R.S. L. & E.

M.R.I.A. F.R.A.S. F.L.S. F.Z.S. Hon. F.C.P.S. &c &c.

ASSISTED BY

EMINENT LITERARY AND SCIENTIFIC MEN.

Natural History.

DESCRIPTIVE AND PHYSIOLOGICAL

BOTANY.

BY THE

REV. J. S. HENSLOW, M. A.

PROFESSOR OF BOTANY IN THE UNIVERSITY OF CAMBRIDGE.

LONDON:

PRINTED FOR

LONGMAN, REES, ORME, BROWN, GREEN, & LONGMAN,

PATERNOSTER-ROW;

AND JOHN TAYLOR,

UPPER GOWER STREET.

1836.

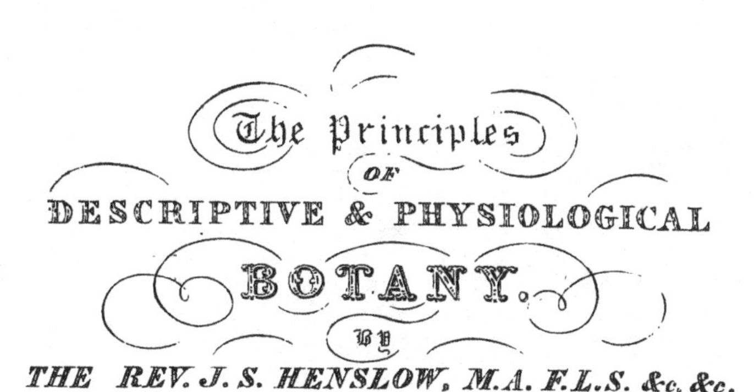

PROFESSOR OF BOTANY IN THE UNIVERSITY OF CAMBRIDGE

H. Corbould del. *E. Finden sculp.*

London;
PUBLISHED BY LONGMAN, REES, ORME, BROWN & GREEN, PATERNOSTER ROW
AND JOHN TAYLOR, UPPER GOWER STREET
1836.

THE

CABINET

OF

NATURAL HISTORY.

CONDUCTED BY THE

REV. DIONYSIUS LARDNER, LL.D. F.R.S. L. & E.

M.R.I.A. F.R.A.S. F.L.S. F.Z.S. Hon. F.C.P.S. &c. &c.

ASSISTED BY

EMINENT SCIENTIFIC MEN.

DESCRIPTIVE AND PHYSIOLOGICAL

BOTANY.

BY THE

REV. J. S. HENSLOW, M. A.

PROFESSOR OF BOTANY IN THE UNIVERSITY OF CAMBRIDGE.

LONDON:

PRINTED FOR

LONGMAN, REES, ORME, BROWN, GREEN, & LONGMAN,

PATERNOSTER-ROW;

AND JOHN TAYLOR,

UPPER GOWER STREET.

1836.

CONTENTS.

INTRODUCTION.

Objects of Botanical Investigation (2.). — Descriptive and Physiological Botany — Sub-divisions (3.). — Advantages of our Pursuit (4.). — Unorganized and organized Bodies (5.). — Distinction between Animals and Vegetables (7.) - - - - Page 1

PART I.

DESCRIPTIVE BOTANY.

SECTION I.

ORGANOGRAPHY AND GLOSSOLOGY.

CHAPTER I.

ELEMENTARY ORGANS AND TISSUES.

External Organs — Conservative and reproductive (9.). — Internal Structure; Elementary Texture; Chemical Composition (12.). — Elementary Organs; Cellular and Vascular Tissues (13.). — Compound Organs — Investing and complex (28.). — Primary Groups or Classes (33.) - 9

CHAP. II.

NUTRITIVE ORGANS.

Fundamental Organs (38.). — Root and Appendages (39.). — Stems (Aërial) (43.). — Internal Structure (45.). — Forms and Directions (53.). — Buds (56.). — Branches (58.). — And their Modifications (61.). — Subterranean Stems and Branches (62.). — Tubers and Bulbs; their Affinity (63.). — Appendages to the Stems (67.) - - - 37

CHAP. III.

NUTRITIVE ORGANS — *continued.*

Leaves, simple and compound (69.). — Vernation (71.). — Forms of Leaves (74.). — Phyllodia (75.). — Transformation of Leaves (78.). — Venation (81.). — Disposition and Adhesion (82.). — Nutritive Organs of Cryptogamic Plants (84.) - - - - Page 59

CHAP. IV.

REPRODUCTIVE ORGANS.

Flower Buds (85.). — Inflorescence — Modes of (86.). — Floral Whorls — Perianth (92.). — Glumaceous Flowers (96.). — Stamens and Pistils (97.). — Disk (101.). — Floral Modifications (102.). — Æstivation (104.) - 79

CHAP. V.

REPRODUCTIVE ORGANS — *continued.*

Fruit — Pericarp (105.). — Forms of Fruit (108.). — Seeds (109.). — Embryo (111.). — Reproduction of Cryptogamous Plants (114.) - 102

CHAP. VI.

MORPHOLOGY.

Abortion (115.). — Degeneration (116.). — Adhesion (118.). — Supernumerary Whorls (119.). — Normal Characters (120.). — Spiral Arrangement of foliaceous Appendages (121.). — Tabular View of Vegetable Organization (129.) - - - - - 116

SECTION II.

TAXONOMY AND PHYTOGRAPHY.

CHAP. VII.

Natural Groups (131.). — Values of Characters (132.). — Subordination of Characters (133.). — Natural Orders (135.). — Artificial Arrangements (136.). — Linnæan System (137.). — Application of it (140.) - - 135

PART II.

PHYSIOLOGICAL BOTANY.

CHAPTER I.

VITAL PROPERTIES AND STIMULANTS.

Vegetable Life (139.). — Properties of Tissues (141.). — Endosmose (144.). — Vital Properties (145.). — Stimulants to Vegetation (152.) - Page 155

CHAP. II.

FUNCTION OF NUTRITION — *Periods* 1, 2, 3, 4.

Absorption (160.). — Ascent of Sap (163.). — Causes of Progression (165.). — Exhalation (169.). — Retention of Sap (172.). — Respiration (173.). — Fixation of Carbon (176.). — Organizable Products — Gum (177.). — Etiolation (179.). — Colours and Chromatometer (182.). — Results of Respiration (189.) - 175

CHAP. III.

FUNCTION OF NUTRITION — *continued* — *Periods* 5, 6.

Diffusion of proper Juice (189.). — Intercellular Rotation (193.). — Local Circulations (195.). — Vegetable Secretions (196.). — Fecula, Sugar, Lignine (197.). — Proper Juices (202.). — Taste and Scent (210.). — Excretions (212.). — Rotations of Crops (218.). — Extraneous Deposits (219.) - - - - - 203

CHAP. IV.

FUNCTION OF NUTRITION — *continued* — *Period* 7.

Assimilation (223.). — Pruning (225.). — Grafting (227.). — Development (230.). — Nutrition of Cryptogamic Plants (233.). — Parasitic Plants (234.). — Duration of Life (235.). — Vegetable Individuals (236.). — Longevity of Trees (239.) - - - 227

CHAP. V.

FUNCTION OF REPRODUCTION — *Periods* 1, 2, 3.

Propagation (243.). — Origin of Flower-buds (245.). — Flowering (246.). — Functions of the Perianth (252.). — Development of Caloric (254.). —

Fertilization (255.). — Formation of Pollen (261.). — Maturation (265.). — Flavour and Colour of Fruit (273.) - - Page 248

CHAP. VI.

FUNCTION OF REPRODUCTION — *continued* — *Periods* 4, 5.

Dissemination (275.). — Modes of Dissemination (279.). — Preservation of Seed (281.). — Germination (283.). — Vitality of the Embryo (290.). — Relation of Bud and Embryo (291.). — Proliferous Flowers (292.). — Hybrids (295.) - - - - - 276

CHAP. VII.

EPIRRHEOLOGY, BOTANICAL GEOGRAPHY, FOSSIL BOTANY.

Epirrheology (298.). — Direction of Roots and Stems (299.). — Botanical Geography (302.). — Fossil Botany (315.) - - - 290

THE PRINCIPLES

OF

DESCRIPTIVE AND PHYSIOLOGICAL

BOTANY.

INTRODUCTION.

OBJECTS OF BOTANICAL INVESTIGATION (2.). — DESCRIPTIVE AND PHYSIOLOGICAL BOTANY — SUB-DIVISIONS (3.).— ADVANTAGES OF OUR PURSUIT (4.). — UNORGANISED AND ORGANISED BODIES (5.). — DISTINCTION BETWEEN ANIMALS AND VEGETABLES (7.).

(1.) Of the advantages which accrue from the cultivation of the natural sciences, sufficient has been said in the treatise of Sir J. Herschel, forming our fourteenth volume; and Mr. Swainson, in his discourse, which forms our fifty-ninth volume, has further exposed the importance of the study of Natural History in general, and more particularly of that department which he so successfully cultivates. In introducing the science of Botany to the general reader, for whom more especially this volume is designed, rather than for the scientific adept, it will be right that we should follow the example which has thus been set us, and say a few words by way of introduction to our present subject. Whenever we are about to enter upon any science which is new to us, it is always advantageous to take a general survey of the limits within which it is restricted, and to obtain some notions of the objects of which it professes to treat. We shall, therefore, offer a few remarks upon the position which Botany holds with respect to other kindred branches of Natural History; and point out the separate and subordinate departments into which it may be advantageously divided.

(2.) *Botany.* — In the most extended sense of the term, Botany may be considered as embracing every inquiry which can be made into the various phenomena connected with one of the three great departments into which the study of nature is divided, and which is familiarly styled the Vegetable Kingdom. And this inquiry should extend as well to the investigation of the outward forms and conditions in which plants, whether recent or fossil, are met with, as to the examination of the various functions which they perform whilst in the living state, and to the laws by which their distribution on the earth's surface is regulated. We may conveniently arrange these several phenomena under two heads. The one may be called the "Descriptive" department of the science, being devoted to the examination, description, and classification of all the circumstances connected with the external configuration and internal structure of plants, which we here consider in much the same light as so many pieces of machinery, more or less complicated in their structure; but of whose several parts we must first obtain some general knowledge, before we can expect to understand their mode of operation, or to appreciate the ends which each was intended to effect. In the "Physiological," which is the other department, we consider these machines as it were in action; and we are here to investigate the phenomena which result from the presence of the living principle, operating in conjunction with the two forces of attraction and affinity, to which all natural bodies are subject.

(3.) *Subordinate departments.* — Each of the two

departments mentioned in the last article admits of subdivision; and the several subordinate departments thus formed become a register of special observations. Thus, the descriptive department will include a "Glossology," or mere register of technical terms — composing a conventional language, by which the description of plants is facilitated, and a comparison of their forms and peculiarities rendered clear and precise, without any periphrasis or unnecessary prolixity. It will also include an "Organography," containing a particular account of the several parts or organs of which plants are composed. A third subordinate department is styled "Phytography," in which a full description of plants themselves is given: and lastly, we have the "Taxonomy" of this science, in which plants are classified in a methodical manner, according to some one or other of those various methods or systems, which serve to facilitate our knowledge of the forms and relations of the numerous species already discovered. We do not, however, propose to treat our subject with so much technicality. In descriptive botany we shall chiefly restrict ourselves to the more general details of Organography, and include in this department whatever we may find it necessary to say on Glossology. The reader may then consult the general index at the end of the volume, whenever he meets with a word which requires explanation, and he will be referred to the page and article in which such explanation is given. Phytography is entirely subordinate to Taxonomy, or Systematic Botany, which forms no part of our scheme, beyond what is necessary to give the reader some general notions of the manner in which plants are described and classified in the most celebrated systems of systematic authors. We shall enter somewhat more fully into the details of Physiological Botany, as this subject possesses a more general interest, owing to the numerous and striking phenomena, of practical and economical importance, which it enables us to explain.

It is more usual, indeed, to restrict the term Botany entirely to the descriptive departments, in which, as might have been expected, and as the nature of the case requires, much greater progress has been made than in the physiological. It is, in fact, only very lately that any successful attempt has been made to connect the numerous facts which have been long accumulating relative to the various phenomena which attend, and the laws which regulate, the functions performed by the living vegetable.

(4.) *Advantages of our pursuit.*—The old and by-gone sneer of "*cui bono,*" by which the naturalist was formerly taunted, now offers no serious impediment in the way of those who are willing to inquire for themselves. Even the few who still think that no advantage would result from the encouragement of natural history as a branch of general education, no longer attempt any very decided opposition wherever they meet with others prepared to uphold it. Our pursuit has been so often and so satisfactorily shown to be productive of direct practical benefit to the general interests of society, that nothing further need here be said on that topic. But we would more especially recommend it as a resource which is capable of affording the highest intellectual enjoyment; and as much worthy of general notice for mental recreation, as air and exercise are for our bodily health. All who feel an unaccountable delight in contemplating the works of nature; who admire the exquisite symmetry of crystals, plants, and animals; and who love to meditate upon the wonderful order and regularity with which they are distributed; possess a source of continued enjoyment within themselves, which is capable of producing a most beneficial effect upon their temper and disposition, provided they do not abuse these advantages by making such studies too exclusively the objects of their thoughts and care. Above all, they must beware of pampering the ridiculous ambition of surpassing others in the extent of their collections, or of fostering an absurd and captious jealousy about maintaining the priority of their claim to this or that particular observation or discovery. We do not go so far as some persons, who seem inclined to believe that these pursuits are of themselves capable of producing a decided improvement in our moral sensibilities; but we hail that joy which is felt in the pursuit of such occupations, as a sacred gift, which may be compared to the rain from heaven, sent for the benefit of all: for increasing the temporal welfare both of the just, and of the unjust: for procuring blessings equally to the good and to the evil; but which the former only know how thoroughly to appreciate, and to apply to the highest and best advantage.

Botany has its peculiar interest, from embracing the study of natural bodies which form the connecting link between the animal and mineral kingdoms. If plants ceased to grow, animals would cease to exist. No animal derives its food immediately from unorganised matter; and though there are many which prey upon other animals, yet the victims have always been themselves nourished by some plant. Nothing can exceed the wonderful manner in which provision is made for the constant supply of those myriads of animated beings which people the earth, ocean, and atmosphere. Most of them are not content with every chance vegetable that may be growing in their path; and many are to be fed, and can only be fed, upon some one or two kinds of vegetable, and would inevitably starve upon every other besides! When, then, we seek to investigate the laws by which the distribution and the very existence of animals is regulated, it is of consequence that we should not overlook even the minutest moss or fungus that we can detect. It is by such plants that the first step must often be made towards rendering the barren and desolate rock a fertile and productive soil, and converting a spot apparently destined to eternal silence into a scene of lively bustle and delight.

(5.) *Unorganised Bodies.*—The most prominent distinction that subsists between the various natural bodies that surround us, is derived from their possessing or being destitute of an organised structure. The want of organisation is the peculiar characteristic of mere brute matter, and affords an evidence of the absence of the living principle; and is a clear proof that it has not been present in those bodies during their formation or increase. On the other hand, the slightest trace of organisation discoverable in any natural body is a complete proof that life is, or at least was once, present in that body. The separate particles of which unorganised bodies are composed, are either elementary atoms, or compound molecules, in which certain elementary atoms are united together by the force of affinity in a definite proportion. When these separate particles, or "integrant molecules" as they are termed in mineralogy, are allowed gradually to coalesce from a state of solution or of fusion, they then arrange themselves into various regular geometric forms, called crystals. These crystals can increase in size only by a further juxtaposition of similar molecules added to them *externally*. When the peculiar circumstances under which they may be placed do not allow these integrant molecules to arrange themselves into crystalline forms, they may still be able to combine together into shapeless masses, which possess the same homogeneity of character as though they had been regularly crystallised. All such combinations of unorganised matter are termed "simple minerals." Compound minerals, such as rocks and stones, the ocean, the atmosphere, are merely heterogeneous admixtures of simple minerals, which naturally exist under a solid, liquid, or gaseous form. When aggregated into large masses, these "compound minerals" constitute our earth, and probably also all the various heavenly bodies.

(6.) *Organised Bodies.* — Although organised bodies are made up of the same elementary atoms as those which compose unorganised bodies, yet are they distinguishable from these latter, not merely by the presence of the living principle, but completely and satisfactorily by the manner in which they increase. The various parts or organs of which such bodies are composed are not homogeneous in their structure, like those of simple minerals; and their increase is effected by an assimilation of certain particles adapted to its growth, which are received into the system through certain cavities, or vessels, from whence they are elaborated, by a peculiar process, into specific compounds, adapted to the nutrition and development of the individual. These effects depend upon the presence and activity of a distinct force, peculiar to the condition under which organised matter exists, viz. that mysterious principle which we call "life," — a something totally different in its mode of action from any of the forces to which unorganised bodies are subjected; and capable of controlling, and, to a certain extent, of counteracting, the effects of those forces. One striking peculiarity in the vital force is its variable condition, and ultimate secession from all organised bodies whatever. However effectual, for a time, in counteracting the influences of the two other great forces of nature, attraction and affinity, a period, sooner or later, does always arrive, in which it ceases to operate, and abandons to silence and inactivity the dust and ashes which it had for a little while collected, and employed in forwarding the high interests of animated nature.

(7.) *Animals and Vegetables.* — We may distinguish organised bodies into animals and vegetables; and our daily experience is sufficient to satisfy us of the propriety of such a division. Yet is it extremely difficult, and has hitherto baffled the attempts of naturalists, to point out the precise limits which separate these two kingdoms of organised nature; and no definitions of what is a plant, and what is an animal, have yet been framed sufficiently guarded and precise to satisfy all the conditions under which different organised bodies are found; but, to this day, there are some objects which it is very doubtful under which class they ought to be arranged. Among the higher tribes of organised bodies, indeed, there is no difficulty in pointing out numerous lines of demarcation between the two kingdoms; but, as we descend in the scale of each, we find an increasing similarity in external characters, and a closer approximation between the analogies existing in many of those functions which mark the presence of the living principle, both in the animal and in the vegetable kingdoms. Perhaps, until the contrary shall be distinctly proved, we may consider the superaddition of "sensibility" to the living principle as the characteristic property of animals; a quality by which the individual is rendered conscious of its existence or of its wants, and by which it is induced to seek to satisfy those wants by some act of volition. It has been supposed—and both analogy and experiment appear most fully to confirm the supposition—that a sense of pain is very nearly, if not entirely, absent in the inferior tribes of animals. Even in the higher tribes, certain parts of the body are incapable of receiving pain; and there seems to be no absurdity in considering that an animal may be endowed with just so much sensibility as may be sufficient to prompt it to select its food, though at the same time its body may be so organised as to be incapable of transmitting painful sensations. But the most constant, if not universal, distinction,—and one which we can readily appreciate, between animals and vegetables,—consists in the presence or absence of those internal sacs or stomachs, with which the former alone are provided, for receiving their food in its crude state, previously to its being elaborated by the organs of nutrition.

PART I.

DESCRIPTIVE BOTANY.

SECTION I.

ORGANOGRAPHY AND GLOSSOLOGY.

CHAPTER I.

ELEMENTARY ORGANS AND TISSUES.

EXTERNAL ORGANS — CONSERVATIVE AND REPRODUCTIVE (9.). — INTERNAL STRUCTURE ; ELEMENTARY TEXTURE ; CHEMICAL COMPOSITION (12.). — ELEMENTARY ORGANS ; CELLULAR AND VASCULAR TISSUES (13.). — COMPOUND ORGANS — INVESTING AND COMPLEX (28.). — PRIMARY GROUPS OR CLASSES (33.).

(8.) *Organs.* — THE various parts of which a plant is composed have been called its " organs ;" and this term is equally applied to those external portions, which may readily be recognised as being subordinate to the whole, such as its leaves, roots, flowers, &c., as to certain minute cells and vessels, of which its internal structure consists. De Candolle has included every inquiry, both into the external and internal organisation of plants, under the title of " Organography ;" although such details as belong to their external characters have a more exclusive reference to our descriptive department, whilst those which relate to their internal organisation are more especially introductory to our physiological.

(9.) *External Organs.*— The principal external organs of which a plant is composed are familiar to every one. They are, the root, stem, branches, leaves, flowers, &c. These organs may be conveniently grouped under two heads, characterised by the nature of the functions which they are severally destined to perform. The root, stem, branches, leaves, and some other appendages to each of these, are concerned in carrying on the function of nutrition, or that act by which the life of every separate individual is maintained; and these are, in consequence, styled the "Conservative" organs. The flower and fruit, with their various appendages, are connected with the function of reproduction, by which the continuance of the species is provided for; and these are, therefore, named the "Reproductive" organs.

(10.) *Conservative Organs.* — The conservative organs, again, may be separated into two series. Every one is acquainted with the fact, that the stems of most plants are above ground, and that they affect a more or less erect position, and are constantly being developed upwards, whilst the roots of most plants penetrate the soil with an evident tendency downwards. An imaginary plane, intersecting the plant at the point whence these opposite tendencies originate, is called the neck: the stem, and the various organs which accompany it, are styled the "ascending," and the root and its appendages the "descending" series. But these definitions do not exactly represent the truth, since there are certain stems which are strictly subterranean, and have a tendency to creep below the surface of the soil; whilst there are also certain roots which are aërial, and some of these scarcely indicate any downward tendency. The terms employed in defining the two series must, therefore, be considered as indicating certain facts, which are very generally, though not universally, applicable to the several organs included under each.

(11.) *Reproductive Organs.*— The reproductive organs may also be classed under two series. The first is the "Inflorescence," which includes the flower and the various appendages to that part of the stem on which it is seated; and the second is the "Fructification," which embraces the seed, and the different envelopes by which it is surrounded, and which collectively are termed the fruit. This latter series, indeed, consists of organs which had previously belonged to the former series during the early stages of their development; but, as a very material alteration takes place in their condition after the flower has expanded and faded, they are considered as having so far changed their character as to merit a different name from that which they before possessed. But here, again, our definitions do not apply to the whole mass of vegetation, since no flowers or seeds are ever produced by the lowest tribes of plants; but they are propagated by little bodies termed "sporules," which do not require any previous process for securing their fertility, similar to that which we shall hereafter show to be essential to the perfection of true seeds.

(12.) *Internal Structure.*— Before we enter more fully into further details respecting these and the other external organs, we propose to examine the internal structure of plants; especially as there are certain investing or cuticular organs, which cannot well be described without referring to the elementary organs, of which the whole structure of the vegetable is composed.

The great simplicity of the vegetable structure, when contrasted with the complexity of that of animals, is very remarkable; and whilst every separate function performed by the latter, seems to require an organ of a peculiar construction, the functions of vegetation are all carried on by the intervention of a few simple tissues of the same kind. Probably, however, this extreme simplicity is much overrated; for as yet we know very little of the numerous slight modifications which different plants exhibit in the arrangement of the several parts of their tissue, and it may be reasonably conjectured, that every modification of this sort, however slight, implies some corresponding alteration in the mode of performing the function. If we cut or fracture any portion of a living plant, we find it to be made up of solid and fluid parts, and with the aid of the microscope we may observe the manner in which these parts are disposed. The solid portions appear somewhat like a spongeous body, penetrated by minute cavities, through which the fluids are dispersed. If we now take a very thin *transverse* slice of some succulent stem, as of a cucumber (*fig.* 1.), and

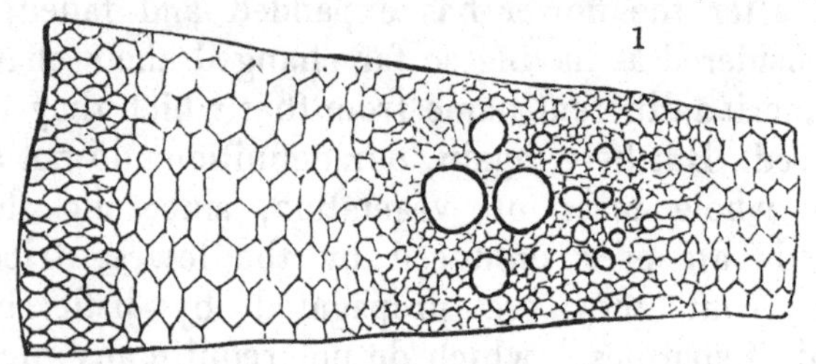

examine a portion of it under lenses of high powers, it will present the form of a distinct network, the meshes of which consist of angular figures, differing in the number of their sides, and in the degrees of regularity with which they are disposed. In some cases the regularity of their form and disposition is very remarkable; and they are frequently hexagonal. The meshes in some parts of the slice are much smaller than in others, especially where they are observed to surround certain circular openings of a different appearance from the rest of the cavities. If another slice be taken *longitudinally* through the stem (*fig.* 2.), and a portion of this be

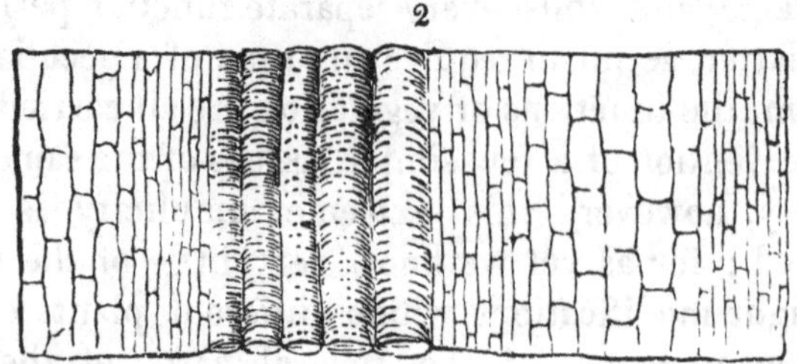

examined in a similar manner, the netlike tissue presents a somewhat different appearance. The meshes are for the most part quadrilateral, or nearly so, and generally elongated in the direction of the axis of the stem. The circular openings observed in the former fig. (1.)

are found to be the sections of tubes, which are often variously marked by dots, lines, and, in some instances, are composed of a spirally twisted filament. These appearances evidently show us that the vegetable structure is composed of polygonal cells and cylindrical tubes, so arranged that they lie with their greatest lengths parallel to the axis of the part in which they are found. Among the lowest tribes of flowerless plants, which form an extensive class, no tubes are observable, and their whole mass is composed of cells alone.

(13.) *Elementary Textures.*—If we now examine the materials of which these cells and tubes are constructed, we find them to consist of a delicate, homogeneous membrane, of extreme tenuity, generally colourless, and without any distinct traces of organisation. Besides this, there is a fine cylindrical fibre, which might be compared to transparent catgut; and this is often spirally twisted and variously ramified upon the surfaces of the cells and tubes, in a manner which we shall presently describe. It is supposed that all the modifications observable in the internal organisation of plants result from the various combinations which take place between these two elementary textures, "Membrane" and "Fibre."

(14.) *Chemical Composition.*—It has not been ascertained whether these two organic elements of the vegetable structure are identical in chemical composition, or whether, indeed, the membrane and fibre which compose the cells and tubes in different parts of plants are always of the same kind. The inquiry would be one of extreme difficulty, if not of absolute impossibility, with the present resources of chemistry. All that is known of the composition of these textures has been derived from experiments made upon the gross material, imperfectly separated from the various matters which the cells and tubes contain. In this state it is found to be composed of the three elements, oxygen, hydrogen, and carbon; but the exact proportion in which these are united is uncertain, if, indeed, it be always the same. In the several products of vegetation—woods, gums, resins, &c.—the proportions between these three elements vary considerably; and even a fourth element, azote, enters as a fundamental ingredient into some of them. It should seem that the atoms which compose the organic molecules in the elementary textures of vegetables, are held together by some vital property, rather than by the laws of chemical affinity; for although these substances may, with certain precautions, be long preserved in much the same state as that in which they were left when the vital principle was first abstracted from them, yet there appears to have been no very definite chemical union between their atoms, which are no sooner abandoned to the influence of surrounding media, than they enter into new combinations distinct from that under which they existed in the living plant.

(15.) *Elementary Tissues.*— There are two elementary tissues, which are respectively composed of the two kinds of elementary organs, the cells and tubes already noticed. The one is called the "cellular" tissue, and consists entirely of cells, and constitutes the chief bulk of every vegetable: the other is the "vascular" tissue, and is made up of tubes; but this latter tissue is found only in certain families of plants. The vascular penetrates the cellular tissue in thin cords, which are composed either of single tubes, or more frequently of bundles of tubes, running continuously throughout the plant, and passing into the leaves, where the tubes separate, and diverge in various directions, and form the veined-like appearance which these organs generally present.

(16.) *Cellular Tissue.*— If a fragment of any plant be allowed to macerate for some days in water, or if it be subjected to the action of nitric acid, the several elementary organs of which it is composed will separate from each other, and may then be examined in an isolated state. When thus detached, the cellular parts are found to have been made up of minute vesicles, or

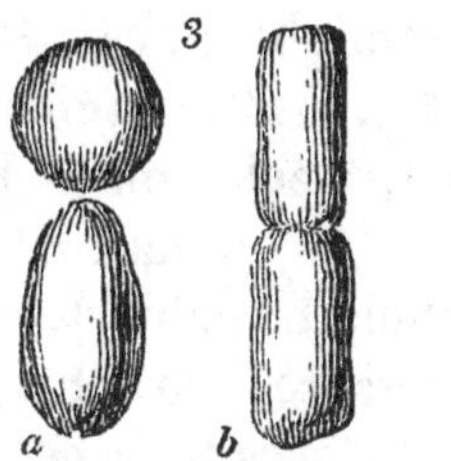

bladders (*fig. 3.*). In some cases these vesicles are nearly spherical (*a*); and, in others, they approach the form of short cylinders (*b*); and in others, again, they are lengthened out, and, tapering at each extremity, present a fusiform or spindle-shaped appearance (*c*). The shortest diameters of those cells which are more or less spheroidal, vary from the $\frac{1}{1000}$ to the $\frac{1}{30}$ of an inch; but are more frequently found between the $\frac{1}{500}$ and $\frac{1}{300}$. The fusiform cells, sometimes termed "closters," which abound in the woody fibre of trees, vary in breadth, at their thickest part, from the $\frac{1}{3000}$ to the $\frac{1}{200}$ of an inch. It is, therefore, entirely owing to the close packing and mutual compression of these vesicles, that they assume a polygonal form in the integral state of the tissue. We may compare the general appearance of this tissue to a mass of froth, obtained by blowing bubbles in soap suds or gum water. The bubbles, by mutual pressure, assume a polygonal structure towards the centre of the mass, but have spherical surfaces towards the outside. In the cells which are thus formed, however, each cavity is separated from its neighbour by only a single partition; whilst, in the vegetable tissue, each partition is of course double. As the cellular tissue alone, without tubes, exists in a large class of plants, it is evident that the most general functions of vegetation must be carried on by it: but, as such an inquiry belongs to the physiological department, we need not say any thing concerning it at present.

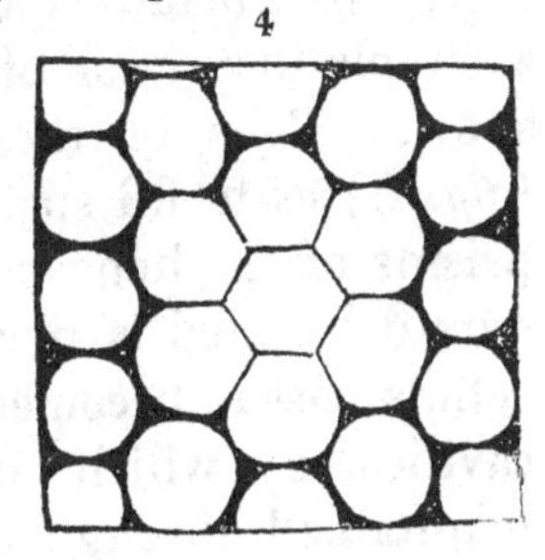

(17.) *Polygonal Structure.*— If we place a number of equal circles in contact, on a plane surface, each circle may be touched by six others; and if we suppose them to be so pressed together, that the curvature of each circle at the points of contact may pass into straight lines, the circles will become hexagons (*fig.* 4.). If a number of spheres, of equal size, be in contact, each may be touched by twelve others (*fig.* 5. *a*); and if the whole be subjected to pressure, so that their surfaces may become flattened at these twelve points, the spheres will become

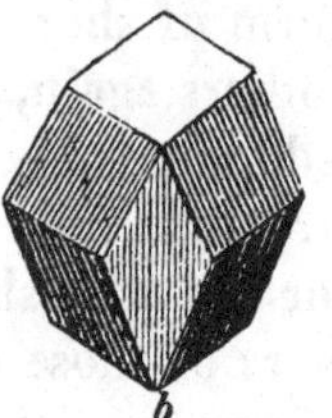

rhomboidal-dodecahedrons (*fig.* 5. *b*). But, as the vesicles which compose the cellular tissue are never exactly of the same dimensions, the polygonal forms which they assume will not be so strictly regular as the geometric figure we have just mentioned. Still, there is often a very marked approximation towards such a regularity; more especially in those parts of the plant which are the best developed, or have been most securely defended, as in the case of the pith, from the influence of disturbing causes. Where the vesicles are elongated, the dodecahedrons assume the character of rectangular prisms, terminated by four-sided pyramids, whose faces replace the angles of the pyramids at various degrees of inclination to the axis (*fig.* 6.). If sections be made through these, by planes parallel and perpendicular to the faces of the prisms, they will exhibit either hexagonal or quadrangular surfaces, according to circumstances, as

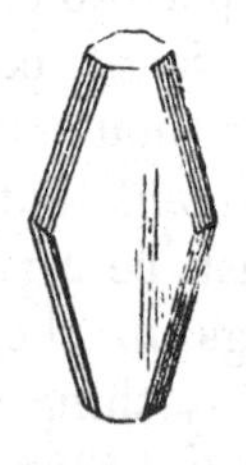

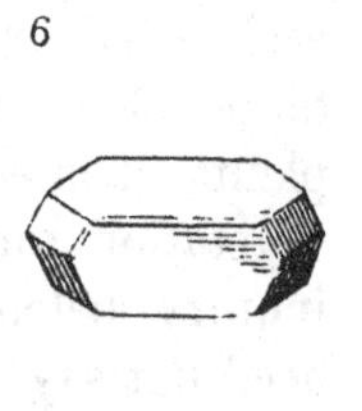

a simple inspection of the diagrams will be sufficient to show. Cells of these forms may be so aggregated (*fig.* 7.) as to fill space as completely as the hexagonal prisms of the honeycomb; but as the extreme regularity here delineated is never actually attained in nature, the cellular tissue becomes every where penetrated by small cavities, by which an intercellular communication is maintained throughout the mass. These channels are termed "intercellular passages," and are very evident in some portions of the tissue, but are not to be detected in others. The forms under which the vesicles appear, upon making a section through the cellular tissue, are much influenced by local pressure, distension, and the more obscure causes which depend upon the specific qualities of each plant; and these forms are detailed with greater minuteness, in works which professedly treat of this part of our subject, in a more elaborate manner than our limits will afford.

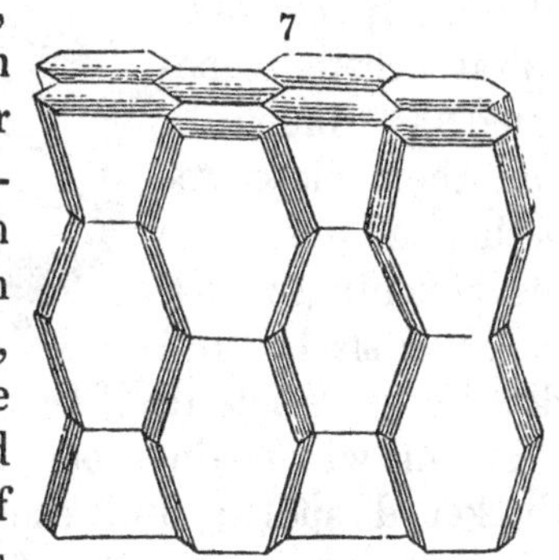

(18.) *Striated and dotted Cells.*— The separate vesicles which compose the cells, frequently exhibit markings upon their surface, whose origin it is not always easy to account for. Many of these appearances were formerly mistaken for open pores through the membrane, by which a communication was supposed to subsist between two contiguous cells. Some observers have considered them to be glands; and others have described them as nascent vesicles, generated within the surface of the old cells, and which are afterwards developed, and thus are formed into new tissue. The best representations of these various appearances, is given by Mr. Slack, in the forty-ninth volume of the "Transactions of the Society of Arts;" and he is inclined to refer the greater part of them to one common origin, viz. the modification of the conditions under which the elementary fibre is developed on the inner surface of the vesicles. In some vesicles, this fibre is spirally coiled over the whole surface, and the contiguous coils are blended together, so as to render it very difficult to distinguish them: in others, the coils are wide apart, and distinctly visible (*fig.* 8. *a*). In some cases the fibre is branched (*b*); and in others, the branches graft together, and the surface of the vesicle then appears reticulated; whilst it sometimes happens, that the coils of a closely developed spiral become separated at intervals, and then close together again, so as to leave openings which look like slashes and dots in the vesicle itself (*c*). There are some cases, however, in which the dots on the vesicles appear to be thickened spots; and especially those which abound on the elongated cells, forming the woody fibre of Coniferous, and some few other trees. These are very peculiarly marked by large dots of a glandular aspect, with a dark spot in the centre (*fig.* 9.); which latter circumstance, however, may probably be owing to the manner in which the light is refracted through them. It is a remarkable fact, that these appearances are strictly imitated in many fossil woods; and botanists are thus enabled, by the inspection of a small fragment of such plants, to pronounce with certainty, upon the Class and Order to which they have belonged. In some cases it happens, that the elementary fibre alone remains entire, like a skeleton to the tissue, whilst the membrane which originally formed the walls of the cells has been obliterated. It is unnecessary to dwell further upon the various appearances which the cellular tissue presents, especially as nothing whatever is known of the way in which a dissimilarity of structure, is connected with any modification in the functions performed by it.

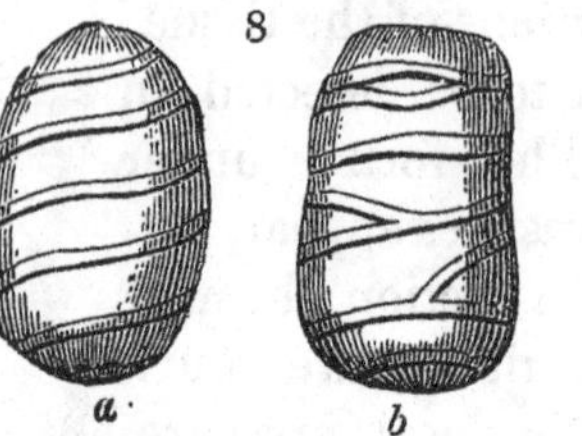

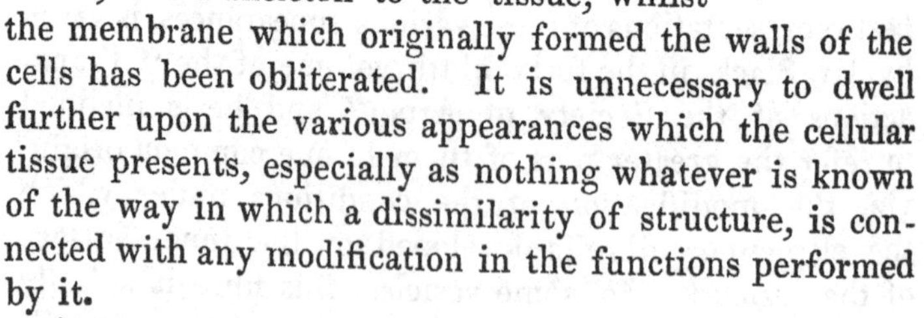

(19.) *Contents of the Cells.* — The cellular tissue is every where replete with juices, containing minute granules of amylaceous, resinous, and other qualities, which appear to be the result of peculiar secretions, formed by the vesicles themselves. Those which compose the woody fibre, secrete an abundance of a carbonaceous material, which ultimately fills them, and gives consistency to the stem. The juicy contents of some cells are highly coloured; and even contiguous cells often contain liquids of different tints, although there is no apparent difference in their structure, which might indicate some cause for such diversity. Indeed, the brilliant hues of flowers, and the various tints of the foliage, all depend upon the coloured juices, or the globules floating in them, which are contained in the vesicles of the cellular tissue, and have been elaborated by them; but they never depend upon the organic membrane itself, of which they are composed, and which is always colourless, or, at best, only slightly tinged with green.

(20.) *Raphides.* — But, besides the strictly organic compounds, there are also certain chemical combinations whose results appear in the form of minute crystalline spiculæ, which have been deposited from the heterogeneous admixture contained in the cells. These have been termed "raphides," and were originally considered to be organised bodies. One of most common occurrence, is the oxalate of lime, the crystals of which are sometimes of such magnitude, and their forms so complete, that the law of their crystallographic structure may be readily recognised.

(21.) *Cavities in the Tissue.* — Besides the intercellular passages mentioned above (art. 17.), there are other well-defined cavities in the cellular tissue, which serve either for the reception of various secreted matters, as resins, oils, &c., or else contain air. The former are termed "receptacles," or "vasa propria," and are commonly of a spheroidal, cylindrical, or oblong form, the result of an enlargement of the intercellular passages, or of a rupture in the tissue itself. The latter are termed "air-cells," or "lacunæ;" and, although these are most frequently very irregular in their form, they are often constructed in a more definite manner than the receptacles, and then consist of extremely regular and well-defined spaces, of hexagonal and other geometric forms. In these cases the cellular tissue is so arranged as to separate the lacunæ from each other, both by vertical and transverse divisions (*fig.* 10.); and the whole is placed round the axis of the stem in a beautiful and symmetrical manner. The stems and leaf-stalks of aquatics are every where filled with lacunæ, and the air contained in them serves the purpose of elevating these parts towards the surface of the water.

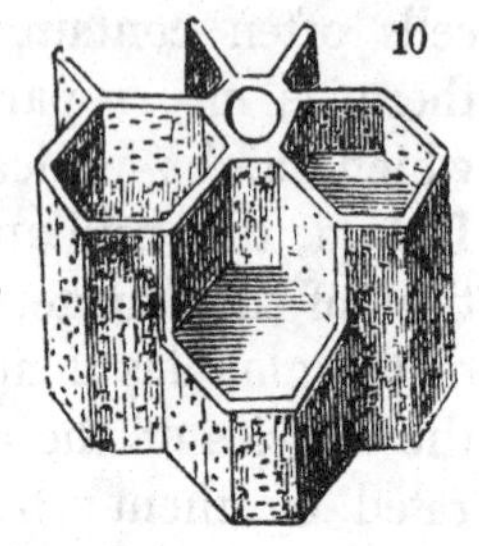
10

(22.) *Vascular Tissue.* — This tissue consists of tubes, which are also formed of membrane, to all appearance identical with that which composes the vesicles of the cellular tissue. Some of these tubes bear a close resemblance to the elongated cells already described, and may certainly be considered as mere modifications of that form of tissue; and, indeed, all tubes, whatever be their length, appear to taper off at each extremity into conical and closed terminations (*fig.* 11. *a*). A communication evidently subsists between some of these tubes, at the point where they overlap each other and are about to terminate, forming an oval perforation of large dimensions. Some tubes are derived from the apposition of cylindrical cells, base to base (*b*), and the subsequent obliteration of the terminal portions of their membrane. In certain cases this membrane remains wholly, or in part, in the form of transverse septa or diaphragms, and then these organs present a tissue intermediate between the cellular and vascular. The true vessels, or long tubes, which more strictly compose the vascular tissue, are distinguishable into two kinds, between which, however, there are cer-

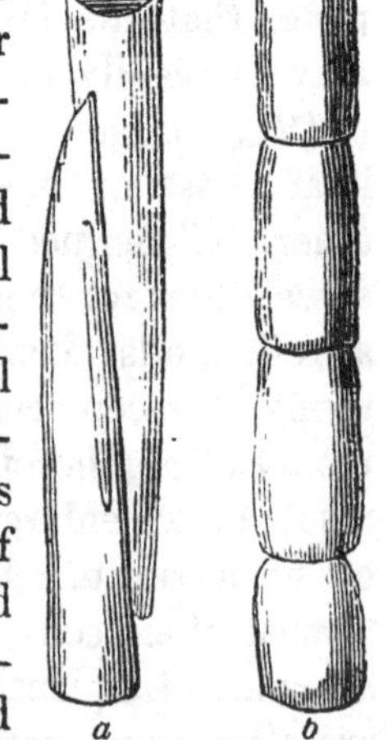
11

tain intermediate forms, which establish the fact of a most intimate connection, and even appear to indicate a common origin. The two kinds of vessels alluded to, are the spiral vessels and the ducts.

(23.) *Spiral Vessels.* — These are generally termed "tracheæ," from the resemblance which they bear to the windpipe, and more especially to the air-cells of insects, which are called by the same name. They consist of a membranous tube, on whose inner surface a cylindrical fibre is spirally coiled (*fig.* 12. *a*); and the whole so completely united, that if the vessel be ruptured, and the thread uncoiled, no trace of the membrane is to be seen, excepting towards the conical extremity of the vessel, where the coils of the fibre are wider apart. In some tracheæ, indeed, the successive coils are not in contact with each other, and then the investing membrane is sufficiently apparent. Sometimes the fibre branches into two threads (*b*), and each continues its course in separate but contiguous coils; and instances may be found, where the contiguous coils of separate threads range (*c*) between this number and twenty-two! The diameters of these vessels vary from the $\frac{1}{3000}$ up to the $\frac{1}{300}$ of an inch. They may be detected with the greatest facility upon tearing asunder the leaves of many plants, and especially are very visible in some species of *Amaryllis*, when they form a set of parallel fibres, nearly as conspicuous as the threads in a spider's web, and are strong enough to support the weight of a considerable portion of the leaf. By carefully unravelling them, they may sometimes be extended to eighteen inches in

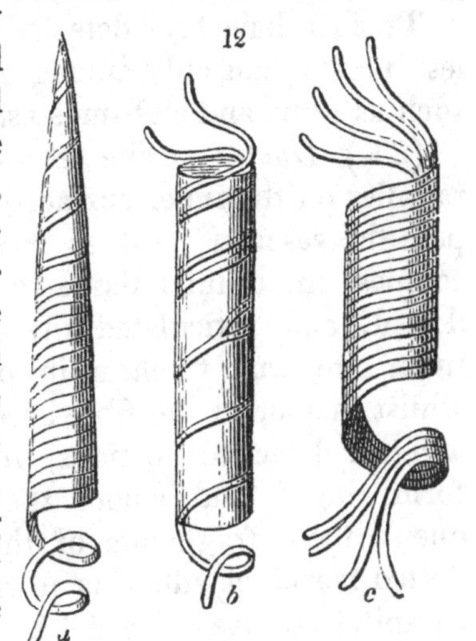

length. When the stems of the Plantain and Banana are cut into slices, the tracheæ in which they abound unravel before the edge of the knife, and form flocculent masses, which may be collected, and wrought into a material possessing certain advantages superior to those of cotton, for the manufacturer. The expense, however, of collecting this delicate substance has been found too great to admit of its being applied to any really useful purpose; as an entire plantain does not yield above a drachm and a half of tracheæ.

Tracheæ have been detected in a very few of the flowerless plants, and only among the higher tribes of them, such as ferns and club-mosses.

(24.) *Ducts.* — The elementary fibre divides and ramifies on the inner surface of some tubes which compose the vascular, just as it does on the vesicles which compose the cellular tissue (art. 18.), and forms linear, dotted, and reticulated markings upon them. Some tubes are true tracheæ in one part of their course, whilst in another the fibre becomes ruptured at intervals, and the detached portions, uniting at their extremities, form rings; and where the ruptures are more frequent, these fragments of the fibre present linear and dotted markings adhering to the surface, and following a spiral course (*fig.* 13.). The name of ducts, is generally given to all varieties of tubes composing the vascular tissue, which are not, strictly speaking, true tracheæ; and they are separately named according to the appearances which the markings on their surface assume; such as dotted, striped, and reticulated ducts. Some authors, however, include all the marked tubes, together with the spiral vessels, under the general

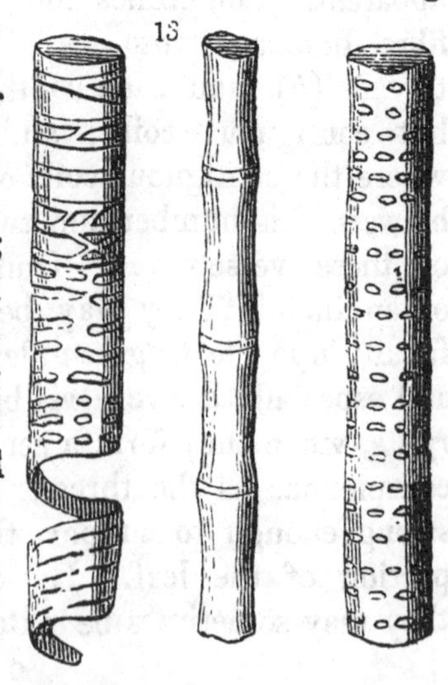

name of tracheæ. The diameters of most ducts are generally larger than those of the true tracheæ, belonging to the same plant; and the dotted ducts, especially, are very distinctly visible to the naked eye, and even large enough to admit of a delicate hair being thrust into them, where they are divided by a transverse section of the stem.

(25.) *Woody Fibres and Layers.*—When a piece of wood is split longitudinally, or in the direction of the stem, it cleaves more readily than when it is broken transversely. And many kinds of wood may be thus split in the direction of the grain, into very thin layers, and these again be subdivided into fibres of extreme tenuity. The fibres obtained by macerating flax, hemp, and other plants used for cordage, are of this description. If these fibres are examined under the microscope, it will be seen that they do not consist of continuous tubes or filaments alone, but are composed of various combinations of vascular and cellular tissue. Every separation in the direction of the fibres (*fig.* 14. *a á*) occasions the disunion of contiguous tubes or vesicles, but any transverse fracture (*b b′*) can be obtained only by the actual rupture of these organs themselves. It is upon this circumstance that the strength of woody fibre depends, which is very different in different plants. It has been experimentally ascertained, that the strength of silk, New-Zealand flax (*Phormium tenax*), hemp, and flax, are respectively as the numbers 34 : $23\frac{4}{5}$: $16\frac{1}{3}$: $11\frac{3}{4}$.

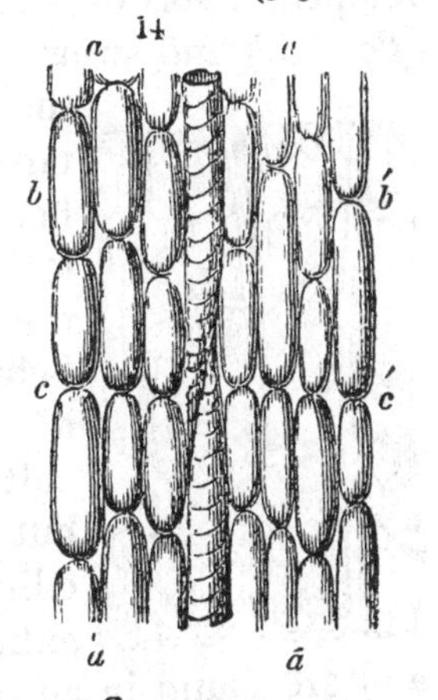

As the cells and tubes are of different lengths, their extremities overlap each other, and thus as it were dovetail the mass together. Wherever a transverse fracture is most readily produced, as in the suture by which a seed-vessel opens, or at the scar which is left where the leaf falls, we may conceive the vesicles which are contiguous to the plane of separation on either side, to be so arranged, that all their ends lie in this plane (*fig.* 14. *c c′*).

(26.) *Contents of the Tubes.*—A considerable diversity of opinion exists as to the probable uses of the vascular tissue in those plants in which it is found. Some observers consider the tracheæ destined to convey air through various parts of the plant; and support their opinion by the fact, that air is very commonly to be observed in them, at least during certain seasons of the year. Others consider all vessels to be channels for the sap and nutritious juices. That most of them contain liquid matter is sufficiently evident, but what may be the precise use of each in particular is at present very uncertain.

(27.) *Vital Vessels.*—Besides the tracheæ and ducts, just described, there is found in certain plants, and possibly in all where the vascular tissue is most developed, a sort of network formed of anastomosing tubes (*fig.* 15.) and situate a little way beneath the surface of the bark, through which fluids certainly pass, in a manner we shall hereafter describe. These tubes are termed "vital vessels," or "ducts of the latex," by their discoverer, M. Schultz. They are by far the smallest of all the tubes, and extremely difficult to be detected in young shoots, but may be seen with tolerable facility as they become older. They are entirely without markings of any kind, and are found in all parts of the plant, from the roots to the leaves.

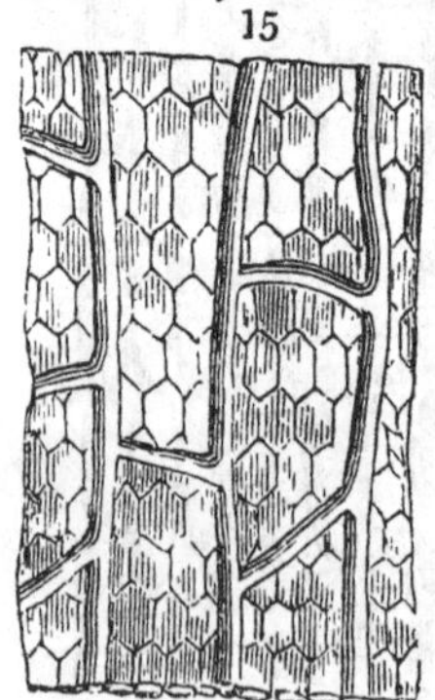
15

(28.) *Compound Organs.*—The organs hitherto described, may be considered as the organic elements out of which plants are constructed, just as we say that minerals are formed out of certain integrant molecules. We have next to notice the various compound organs, which result from different combinations of these elementary organs. These may be considered as of two kinds. The first includes such as are found on the surface of the several external organs, of which in fact they are only subordinate parts, just as the skin, hair, feathers, &c. clothe the body and particular members of animals. We may call these superficial organs, the "Investing organs." The other kind may be styled the "Complex organs," and will include all those which we have already classed under the ascending and descending series, alluded to in art. 10., and of which the investing organs form only subordinate parts.

(29.) *Epidermis.*—The surface of all parts of plants (except the spongioles and some stigmata to be described hereafter) is covered, at least when young, with a thin skin, which may easily be detached, especially from the leaves, and most readily after these organs have been allowed to macerate for a few days in water. This skin is termed the "epidermis," or "cuticle," and when placed under the microscope, it exhibits a delicate network (*fig.* 16.), whose meshes are either

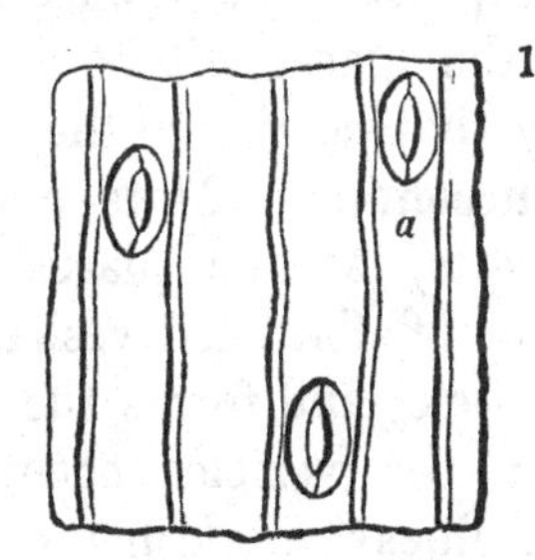

16

either quadrangular, hexagonal, or of other polygonal forms; or else they are irregularly bordered by waved and sinuous lines, extending over the whole surface. Very frequently also, a set of pores may be observed, having a sort of glandular border (*a*), which are scattered over the epidermis at intervals. These pores are termed "stomata." It was not until very lately that the real structure of the epidermis was well understood; but M. A. Brongniart has shown, in the

Ann. des Sciences for February, 1834, that a lengthened maceration causes it to separate into three parts (*fig.* 17.). The outermost of these, consists of an extremely de-

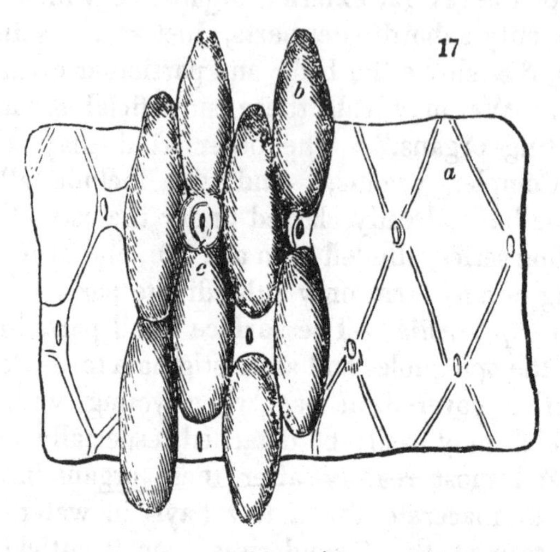

licate homogeneous pellicle (*a*), without any very decided traces of organisation, though occasionally somewhat granulated in its appearance, and also marked by lines, which are merely the spaces left between the impressions made upon it by that portion of the cellular tissue with which it was in contact. It is generally perforated by small oval slits, at the places where the stomata exist. A lamina of flattened vesicles (*b*), is closely united with this pellicle, and forms the second portion of the epidermis; the vesicles occupy the spaces included between the linear markings observed upon the surface. Sometimes this part contains more than one lamina of flattened vesicles. The vesicles are in close contact, excepting immediately under the spaces occupied by the slits in the pellicle. The third part alluded to, consists of the stomata (*c*), which are placed a little further from the pellicle than the lamina of cells last mentioned, and which, as we stated, is in immediate contact with it.

(30.) *Stomata.*—Each stoma is most generally composed of two lunate vesicles (*fig.* 18. *a*), which may be detached by maceration in water, but in the epidermis are in close contact at their extremities, and thus form a sort of border round the area occupied by the slits in the outer pellicle. The space between these vesicles may be contracted or completely closed, by an alteration in their position. Some stomata appear to consist of a single annular vesicle (*b*), which may possibly be occasioned by the blending of two; or this may be owing to an optical illusion. In some cases, the stomata are square (*c*); in others, the orifice appears dark, but whether from the interposition of a peculiar membrane, or merely by the deposit of secreted matters, seems to be doubtful. As the vesicles of the stomata contain granular matter, they appear to be more nearly related to those of the cellular tissue in the substance of the leaf beneath the epidermis, which contain a similar matter, than to the flattened cells which compose this organ itself, and which are generally without grains, and perfectly transparent. Stomata do not occur on flowerless plants, excepting among their higher tribes, and which also possess tracheæ (art. 23.). They are also absent on the submerged parts of aquatics, and are not to be found on certain parasitic plants.

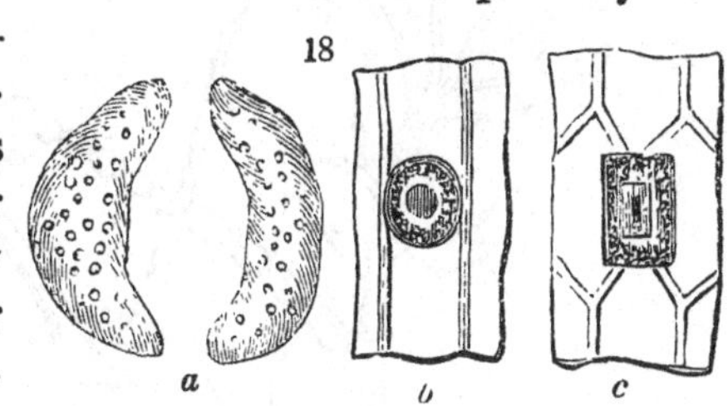

(31.) *Pubescence.*—There are great varieties in the forms under which certain prolongations of the cellular tissue occur, on the surface of different parts of plants. To the naked eye, such appendages to the epidermis resemble hair, silk, bristles, scales, &c., and have received these names in descriptive botany. Under the microscope, they are all found to be composed of cellular tissue; sometimes of a single vesicle, at others of several united (*fig.* 19.). In some, the vesicles are rigid, elongated, and sharp spiculæ; in others they

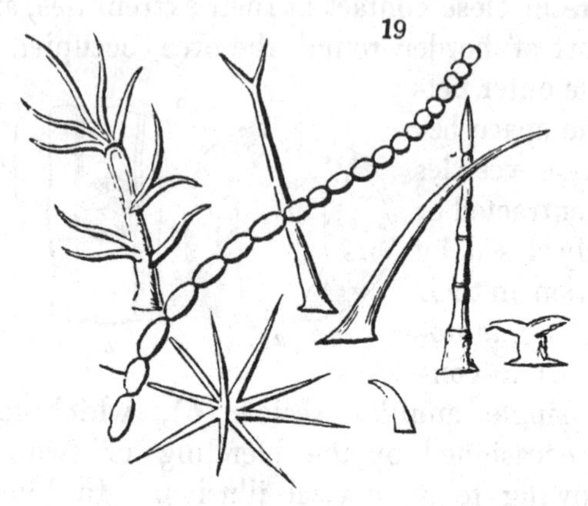

constitute a globular mass of a glandular structure (*fig.* 20.), and secrete various juices of glutinous,

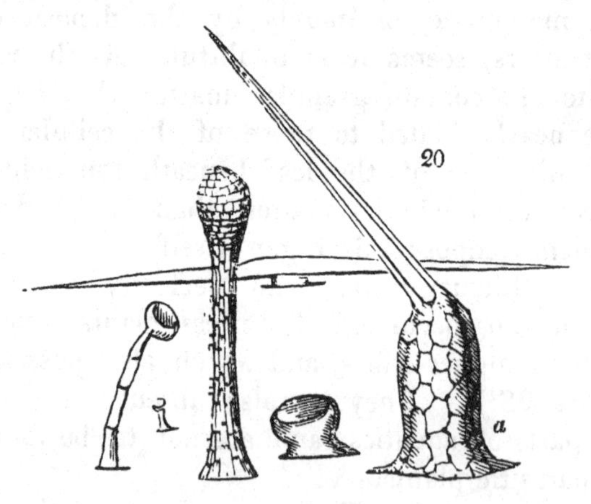

sweet, acrid, and other properties. Stings are sharp-pointed hollow bristles, perforated at the extremity, and seated on a glandular mass of cellular tissue which secretes the poison (*fig.* 20. *a*). When the hand is gently pressed against them, the delicate point penetrates some pore of the skin, at the same time the bristle is forced against the gland at its base, and the poison rises into the tube in a manner strictly analogous to that by which a discharge of venom is effected from the fangs of a serpent's tooth. The bristles have sometimes a stellate form (*fig.* 21. *a*); and sometimes the pubescence is composed of little plates or scales (*b*).

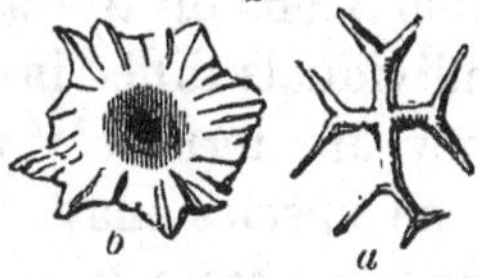

(32.) *Complex Organs.*—Although the epidermis and several of the other investing organs are of a compound character, they are still constructed in a much more simple manner than the organs which they invest. We have proposed, therefore (art. 28.), to separate the latter under the name of "complex organs," which will include all that have been already enumerated under the name of external organs (art. 9.), together with various appendages to be found on some of them. These latter are not so generally noticed by casual observers; but it will be necessary for us presently to describe them, when we treat of the forms and structure of these organs themselves. But we shall here postpone for a while the descriptive details of these organs, in order that the reader may first obtain some general notions of the three great natural divisions under which all plants may be arranged. Although this method of treating our subject may seem to indicate a great want of system, it appears to us highly convenient that every one should be acquainted with these divisions as early as possible before he enters on certain details which cannot be so well appreciated or discussed without an occasional reference being made to them. It must be remembered that we have not proposed to ourselves any very methodical discussion of the several departments of our science, which would have required a series of separate treatises, but that we aim chiefly at conducting the general reader, by such steps as may seem sufficiently adapted to the purpose, to the ready comprehension of some of the best established facts in vegetable physiology, and to give him an idea of what botany proposes to attempt.

(33.) *Primary Groups.*—We apply the term "species" to an assemblage of individuals which have sprung from seeds of the same common stock. Where these individuals differ in certain respects among themselves, they are termed "varieties;" but all varieties of the same species may, under particular circumstances, be produced from the seeds of one plant. When different species bear a striking resemblance to each other, they are classed together in a group which is termed a "genus;" and such genera as agree in several points, form a higher group called an "order;" and those orders which are most nearly related, constitute our chief or primary groups, termed "classes." Minor groups of subordinate value may be formed in each of these; but we do not consider it necessary at present to enter into further details of this kind. We merely propose to explain some of the chief characters by which all plants may be grouped under three distinct classes. The considerations upon which these groups depend, do not rest upon any one solitary fact relative to the structure or functions of all the species they contain; for there is no leading characteristic in either class which is not liable to some objection, if it were to be considered as the only distinguishing mark for deciding the claims of a species to be included in that class. But where one leading characteristic is deficient in one species, and another in another, it is from the aggregate of such as are present that we must decide upon the class to which each should be referred. With very few exceptions, however, nearly all plants may be referred by any botanist, at a single glance, and with unerring certainty, to their proper class; and a mere fragment even of the stem, leaf, or some other part, is often quite sufficient to enable him to decide this question. The names of these three classes are derived from one of the chief characteristics which prevails through *nearly* all the species included under each of them separately. This we shall presently explain; but the reader may understand these names to be *Dicotyledones*, *Monocotyledones*, and *Acotyledones*; and that the two former of these classes have respectively the names of *Exogenæ* and *Endogenæ*. The former names are derived from peculiarities connected with the structure of the seed; the latter, from a consideration of the internal organisation of the plants themselves.

(34.) *Dicotyledones, or Exogenæ.* —

(1.) Structure of the Seed.

Beans, peas, almonds, the kernels of our stone fruits, &c. afford us familiar examples of the structure of the seeds of dicotyledonous plants (*fig.* 22.). When the

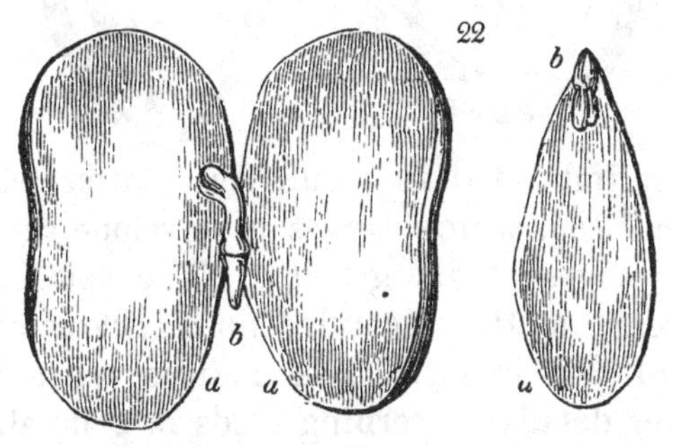

outer skin is removed, we find that they are composed of two large fleshy lobes (*a*), termed "cotyledons," which are attached to a small rudimentary germ (*b*), almost entirely concealed between them. The entire mass forms the "embryo," and the skin which invested it is termed the "seed-cover." After the seed has been sown, and germination has commenced, the two cotyledons expand and represent (what in fact they are) a pair of imperfect leaves, but differ in many respects from the leaves which are subsequently developed. One extremity of the little germ to which the cotyledons are attached, is termed the "radicle," and this descending into the ground becomes the root. The other extremity is termed the "plumule," and consists of the rudimentary leaves and stem. In these examples, where the embryo occupies the whole space within the seed-cover, the fleshy cotyledons contain the nutriment on which the young plant subsists until the root is sufficiently developed to support it. But there are other cases, as in the seeds of the castor-oil plant (*Ricinus communis*), the marvel of Peru (*Mirabilis Jalapa*), &c., where the cotyledons are thinner and more leaf-like (*fig.* 23. *a*), and the embryo is wholly

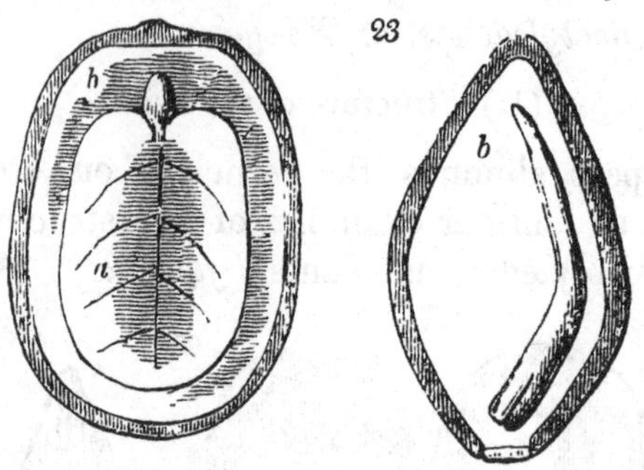

or partially imbedded in a nutritive matter termed the "albumen" (*b*), which serves to develope the plant in the early stages of its growth. The few exceptions which occur in the dicotyledonous character of the embryos of this class, will be noticed when we enter into further details concerning seeds in general.

(2.) Organisation of the Stem.

The most important characters in the organisation of most stems of this class, depend upon the manner in which they increase in thickness. In young and succulent stems, we find a solid cylindric or prismatic mass composed of cellular tissue, and termed the "pith:" this is surrounded by a ring of vessels, consisting of tracheæ and ducts, and named the "medullary sheath." The whole is coated by the epidermis. Afterwards, a further development both of cellular and vascular tissues takes place between the medullary sheath and epidermis, and these form one layer of wood, and also one layer of bark, by the time that a stem of one year's growth is completed. During the second year, a fresh development takes place between the wood and bark previously formed. This fresh matter appears at first as a semifluid or viscous mass, termed "cambium," which is gradually organised, and ultimately separates into two layers — one making an addition to the wood, and the other to the bark, which had been previously formed. Hence a layer of new wood forms a ring round the old wood, and a layer of new bark round the new wood; whilst the old layer of bark, being necessarily thrust outwards, is ruptured and withers, though it still continues to form an outer coat over the whole stem. A layer of fresh wood and another of fresh bark are in this way deposited every year; and in many cases, we may ascertain the exact age of a tree by the number of the concentric zones observable upon making a transverse section of its stem. Thus, in *fig.* 24., *a* is the pith, *b* represents three layers of wood, and *c* an equal number of layers of bark. Besides these concentric zoned appearances on the surface of the section, there are also other traces running in straight lines, radiating from the centre to the circumference, which are formed of cellular tissue, and termed "medullary rays." Either of these three circumstances, then — the existence of a pith, the appearance of concentric zones, or the presence of medullary rays — affords a sufficient characteristic by which we recognise the structure of dicotyledonous plants. The plants of this class are further named "Exogenæ," from the circumstance of their stems increasing in thickness by fresh materials, which are arranged "externally" with respect to the old layers. The oldest and hardest parts of such stems lie towards the centre, as may be readily seen in any tree growing in our temperate zone.

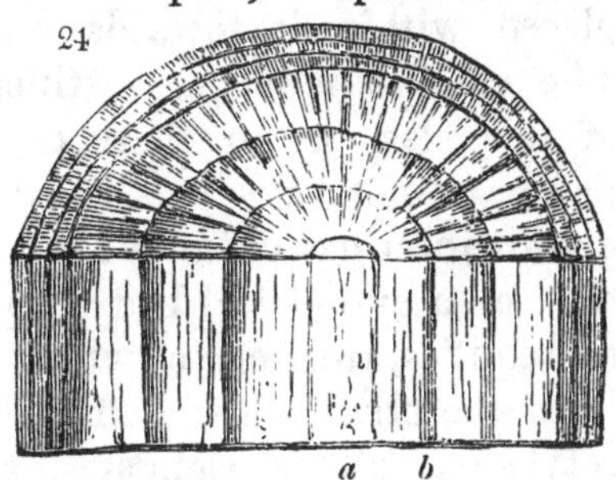

(35.) *Monocotyledones, or Endogenæ.* —

(1.) Structure of the Seed.

The general structure of the seeds of this class may be exemplified by an examination of a grain of Indian corn, wheat, &c.; or of a seed of an onion, lily, &c. (*fig.* 25.). An albuminous mass (*a*) forms the main

25

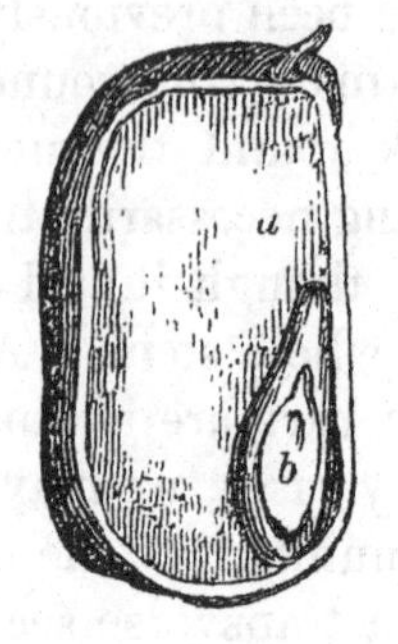

bulk of most of these seeds, and the embryo (*b*) is placed within it towards the centre, or on one side. The embryo is not so distinctly developed in the seeds of this class as in those of the last, and its several parts cannot always be readily recognised before germination has commenced. Its general character is that of a cylindrical body, tapering more or less at the extremities, from one of which protrudes the radicle, and from the other arises a single, conical, and almost solid cotyledon. This elongates, and is ultimately pierced by a leaf, rolled into a conical form, and which was at first completely invested by the cotyledon.

(2.) Organisation of the Stem.

In Monocotyledones, there is no distinction between pith, wood, and bark; but their stems consist of a cylindrical mass of cellular tissue, through which bundles of vascular tissue are distributed in a scattered manner (*fig.* 26.). Every fresh development of new matter is carried towards the centre of the stem, and, as the stem elongates, the outer parts become more and more solidified, whilst the inner remain soft.

26

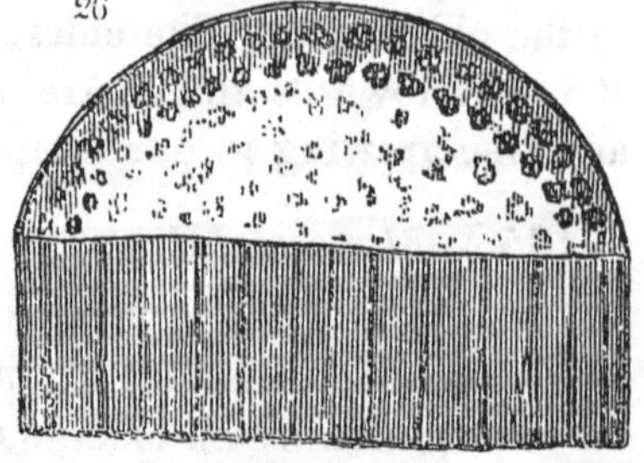

These stems possess no traces of medullary rays. The

plants of this class are termed "Endogenæ," from the circumstance of the newly formed materials being always developed towards the innermost part of their stems. A piece of cane is a familiar example for illustrating this structure; but we have no woody plants in our climate belonging to this class, and very few even which possess herbaceous stems, if we except the hollow culms of the grasses, where the development of the materials towards the centre is not sufficiently rapid to keep pace with the elongation of the stem, and the tissue is in consequence ruptured.

(36.) *Acotyledones.* —

(1.) Structure of the Sporules.

The class to which we now refer, includes an extensive series of plants, grouped under several orders, which differ considerably in many particulars. The whole agree, however, in the important circumstance of never bearing flowers, like those of the two former classes: hence they are termed "cryptogamic," in contradistinction to "phanerogamic," which is applied to all flowering species. Having no flowers, they produce no true seeds; but, in lieu of them, are furnished with what certainly bear a considerable resemblance to seed, viz: small minute granular bodies capable of becoming distinct plants. The manner in which these "sporules," as they are termed, are produced, is very various in the different orders of this class, but forms no part of our present inquiry. They are also variously shaped, but generally spherical or spheroidal, and are not separable into distinct parts, with radicle and cotyledon, like the seeds of phanerogamous plants. In germinating, the sporules are developed by an increase of cellular tissue, which appears in the form of rounded masses and filamentous chords (*fig.* 27.). Among the higher tribes, roots are afterwards produced; and a part which is more or less elevated above the soil, is the representative

27

both of the stem and leaves of phanerogamous plants combined. In the lower tribes, however, there is seldom any separation of parts into distinct organs, but the functions of nutrition are carried on in an obscure manner by the general mass.

(2.) Internal Organisation.

The internal organisation of acotyledonous plants, is not sufficiently uniform in the different orders, to allow of their being characterised by any appellation derived from their mode of development, as in the case of the Exogenæ and Endogenæ. But acotyledonous plants may be separated into two groups: the one, termed "Ductulosæ," characterised by the existence of a vascular tissue, and by a mode of development much resembling that of the Endogenæ; the other, termed "Eductulosæ," or "Cellulares," is entirely composed of cellular tissue. De Candolle even considers the former group, in spite of their cryptogamic character, to possess a monocotyledonous mode of development in the germination of their sporules, and keeps them separate from the others, as a distinct class. The latter group may be strictly termed "Cellulares," from their being composed of cellular tissue alone, and thus separated from the "Vasculares," which will include the rest of vegetation (as well cryptogamic as phanerogamic), possessing a vascular structure. The class Acotyledones is, however, very readily recognisable by its external appearance alone; and the general characters of the several orders which it embraces — ferns, mosses, lichens, seaweeds, fungi, &c. — are pretty familiarly known as examples.

(37.) *Tabular View.*—In the very slight sketch here given of the primary groups under which all plants may be arranged, we have not pretended to notice many terms which different botanists have applied to them; but we shall now collect the substance of what we have said in the form of a table, which may serve to assist the memory of the reader in fixing any of the terms here employed, which may chance to be new to him.

Primary Groups, characterised by certain Considerations taken from particular Parts.

	Embryo.	*Structure.*	*Fructification.*
1.	Dicotyledones.	Exogenæ.	Phanerogamæ.
2.	Monocotyledones.	Endogenæ.	
3.	Acotyledones.	Ductulosæ.	Cryptogamæ.
4.		Cellulares.	

CHAP. II.

NUTRITIVE ORGANS.

FUNDAMENTAL ORGANS (38.). — ROOT AND APPENDAGES (39.). — STEMS (AËRIAL) (43.). — INTERNAL STRUCTURE (45.). — FORMS AND DIRECTIONS (53.). — BUDS (56.). — BRANCHES (58.) — AND THEIR MODIFICATIONS (61.). — SUBTERRANEAN STEMS AND BRANCHES (62.). — TUBERS AND BULBS; THEIR AFFINITY (63.). — APPENDAGES TO THE STEMS (67.).

(38.) *Fundamental Organs.* — WE may refer back to articles 8, 9, &c. for a general notice of the complex organs, which we are now about to describe more in detail, though we do not propose to enumerate all the varieties of form which these organs assume. There are certain appendages both to the stem and root, (or ascending and descending "axes" of vegetation), which are of very little importance in carrying on the function of nutrition. These appendages, as the thorns, scales, tendrils, &c. found on some stems, have without doubt their respective uses; but as the plant may be deprived of them, and still continue to vegetate as freely as when they were present, they are evidently not to be considered as fundamentally essential to the support of life. Moreover, they may in all cases be referred to certain modifications and metamorphoses, which have taken place in one or other of the three organs — the root, stem, and leaf,—which are more especially considered to be the "fundamental organs" of nutrition. The presence of neither of these can be dispensed with without injuring vegetation, and ultimately involving the destruction of the individual; unless where some means have been provided (as we shall see in the case of parasitic plants) to supply their deficiency, or where (as in the lowest tribes of cryptogamic plants) they are probably so blended and confounded together that we are not able to distinguish them.

(39.) *Root.* — The most common position for the roots of plants, is at the base of the stem, from whence they descend into the ground, gradually tapering to a point, and giving off filamentous branches on all sides, in an irregular and indeterminate manner. These branches of the roots are termed "fibrils," and are composed of ducts and cellular tissue, and covered by an epidermis, except at their extremities where the cellular tissue is exposed. It is here that the true absorbents of the root exist, termed its "spongioles." The structure of the main trunk, "caudex," or "tap" of the root (when well developed) is strikingly analogous to that of the stem, except that in dicotyledonous plants there is no pith, and in all cases the epidermis is without stomata. The medullary rays, however, are present; and the bark generally bears a much larger proportion to the whole mass, than in the stem. This latter circumstance is owing to its being kept moist by its underground position, which renders it more capable of distention. In the carrot, this is well exhibited by a difference in the colours of these parts. The concentric woody layers are not distinguishable, and it very seldom happens that tracheæ are found in roots. They are very rarely of a green colour, excepting some of those which are developed above ground; and even then it is seldom more than the spongioles which are thus partially tinted. Where the root has no descending caudex, which in some plants soon dies away, the fibrils are given off from below the neck, or from a flattened disc which represents the caudex, as, for instance, in the bulbs of hyacinths. Roots, however, may be developed from any part of the stem and branches, if these are duly subjected to the influence of moisture and shade; and some plants of tropical climates constantly produce roots from their stems and branches, which descending into the ground become fixed, and serve to support the superincumbent vegetation, and thus enable it to extend over a large tract of ground. The most celebrated example of the kind is the banyan-tree of the East Indies (*fig.* 28.). In this case, it appears that when

28

the roots have reached the ground, the exposed portion assumes the character of a stem. It has, indeed, been asserted that the stem and root are so entirely distinct, that the latter is never capable of assuming the character of the former. But it is not uncommon to find ash-trees which have grown on the stumps of pollard willows and have sent their roots through the decayed wood into the ground; the exposed roots of the ash, when the willows have fallen to pieces, become coated with a green bark, and do not appear to differ in any respect from the trunk itself. At all events, many roots are as capable of producing stems or branches, as these are of forming roots: this is often the case with the white poplar, and certain elms which throw up their numerous suckers, to the great detriment of the pasturage when planted in meadow land.

Besides the important purpose which the root is more especially destined to serve, of absorbing nutriment, it is generally so placed as to take firm hold in the ground, and thus enables the plant to maintain its position in one and the same spot during its lifetime. There are, however, certain plants, as the common duckweeds (*Lemnæ, fig.* 31. *b*), which float on the surface of ponds, whose roots are suspended in the water without ever reaching the bottom. There are others termed "air-plants" (some of the *Orchideæ*), whose roots cling closely to the branches of trees, and derive their nutriment from the moist atmosphere perpetually hanging over a tropical forest; and these plants could not live long if they were planted in the ground.

(40.) *Forms of Roots.* — The various forms which roots assume need not be dwelt upon here; they are such as may be readily learnt in any elementary work, but their description would involve us in details for which we have not space.

(41.) *Appendages to the Root.*—There are not many distinct appendages to be found on roots. In some fibrils, there are swollen nodosities (*fig.* 29.), and on

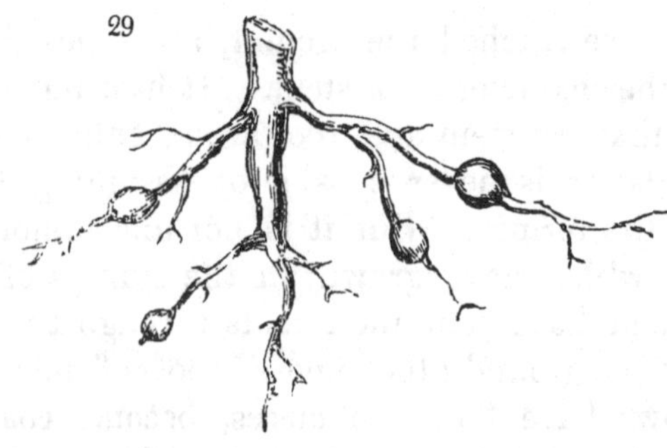

others there are little tuberous excrescences. In some, the fibrils become very fleshy, and are swollen into masses (*fig.* 30.), having an ovate (*a*), palmate (*b*), or fasciculate (*c*) appearance, as in many of the Orchideæ. All these swollen portions serve as reser-

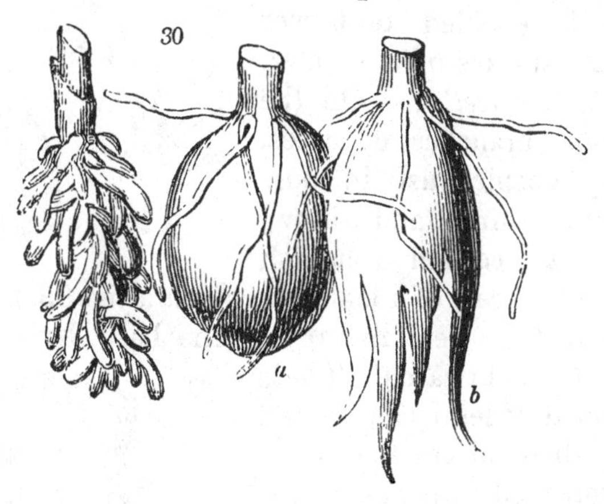

voirs of nutriment for the future use of the plant, but they should not be confounded with certain analogous modifications of the underground portions of stems, which we shall describe when we speak of the real "tuber."

The extremities of some aërial roots, as in the Pandanus, are coated by exfoliations of the epidermis; and the same may be observed on those of the hyacinth. The little *Lemnæ*, or duckweeds (*fig.* 31. *b*), whose roots hang suspended in the water, have a distinct cup-like appendage attached to their extremities. In the early state of their development this formed a membranous sheath (*a*), which completely enveloped them, but which became ruptured at the base as they elongated, and was then carried downwards as they continued to grow.

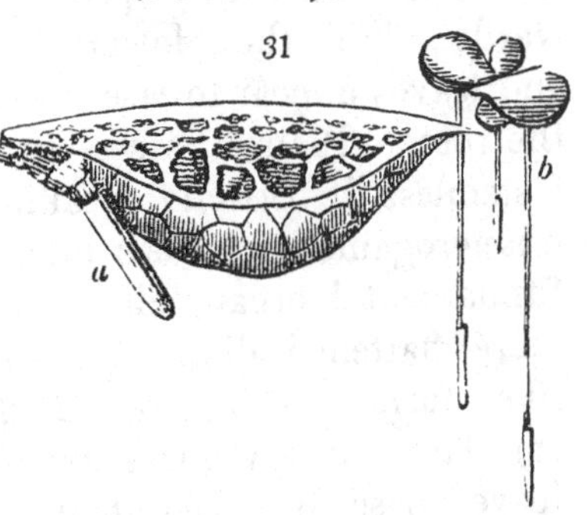

(42.) *Bladders.* — The roots of certains aquatics belonging to the genus Utricularia, are furnished with appendages, in the form of little membranous bladders (*fig.* 32.) which are partially filled with air, and serve to float the plant, in order that it may be enabled to flower above the surface of the water.

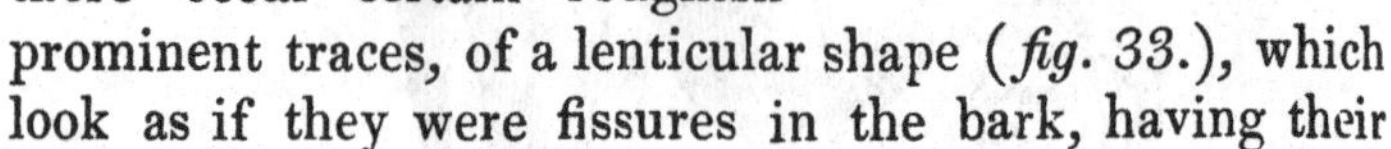

32

(43.) *Lenticellæ.* — On the stem and branches of trees, and very conspicuously in those of the alder, birch, and willow, there occur certain roughish prominent traces, of a lenticular shape (*fig.* 33.), which look as if they were fissures in the bark, having their edges turned outwards. These are termed " lenticellæ ;" and it is at these places that roots are protruded whenever the stem is placed under circumstances calculated to give rise to them.

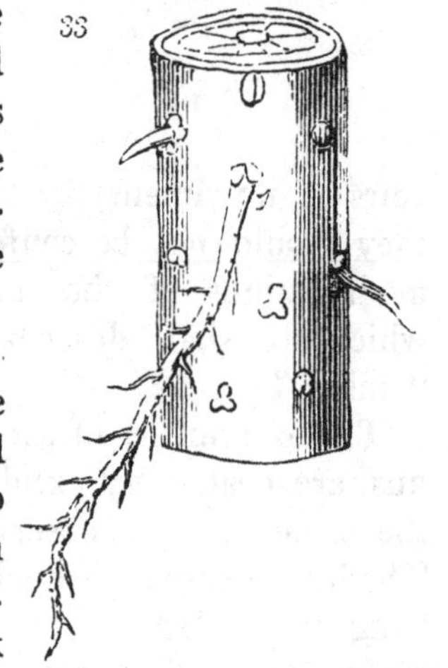

33

(44.) *Stems.* — As the caudex, or main trunk of the root, is not much extended downwards in many plants, so there are many stems which are never much developed upwards; but the flower-stalk and leaves appear to rise immediately from the crown of the root. Plants possessing this character are called "stemless." Strictly speaking, however, there are no phanerogamous plants which are entirely without this fundamental organ, although it is often reduced to a mere flattened disc. Occasionally it assumes a bulb-like form, as in the Cyclamens (*fig.* 34.), where the large woody mass from whence the flowers and leaves arise, is a true stem. In some plants, the stem is wholly beneath the surface of the ground, forming the " subterraneous stem," or " rhizoma ;" but most frequently it rises above it, and composes " the aërial stem," which is called a " trunk," " culm," &c. according to its structure.

(45.) *Aërial Stems.*— The stem is said to be "herbaceous," when it continues soft, and lasts only for a short time; dying soon after the flower has expanded, and the seeds ripened. It is called "woody," when it continues to increase for several years. Herbaceous stems belong to "annuals," "biennials," and "perennials," which are thus named, according to the several periods which their roots continue to live. Woody stems are confined to shrubs and trees; the former having many stems rising from the surface of the ground, and the latter possessing one main trunk, which branches or not, according to the nature of the species to which it belongs. An "undershrub," is where the branches are partly woody and partly herbaceous, so that a portion only dies back every year. Besides these, there are the "succulent" stems, so called from the highly developed state of their cellular tissue, which often remains replete with juices for many years, without hardening into wood.

34

(46.) *Internal Structure of Stems and Roots.*— In arts. 34, 35. we have given an account of the leading differences, observable in the internal composition of the stems of dicotyledonous and monocotyledonous plants; and we have now to explain a few more particulars respecting them.

(47.) *Dicotyledonous Stems.* — In some stems of dicotyledonous trees it is difficult, and in others impossible, to distinguish any separation of the wood into concentric layers. This is especially the case with trees of tropical climates, where vegetation is not liable to the periodic checks which it receives in colder regions. In a few examples, also, the medullary rays are not clearly distinguishable, but the pith and bark are never wanting.

(48.) *Pith.* — The vesicles of the pith are larger and more regularly arranged than those of other parts. It continues to increase in diameter as long as it remains succulent, and in some trees, as the elder, it becomes more than half an inch thick; but generally it is much smaller. After it has lost its succulency and become a dry spongy mass, it scarcely diminishes in size; but where the branch is much distended, the pith is ruptured, and in some cases appears to be nearly obliterated. The stems then become hollow, as in many umbelliferous plants. It always forms a continuous mass through the whole stem; but in some cases it is so much condensed and hardened as to resemble wood at the places where the leaves are attached, as in the horse-chestnut.

Although it is generally without any fibres of vascular tissue, such are found in some plants, as in the elder, where they may be seen, in a transverse section, forming a circle of red dots, a short distance within the medullary sheath. In *Ferula communis* there are so many of these dispersed through it, that the stem has the appearance of belonging to a monocotyledonous plant (*fig.* 35.).

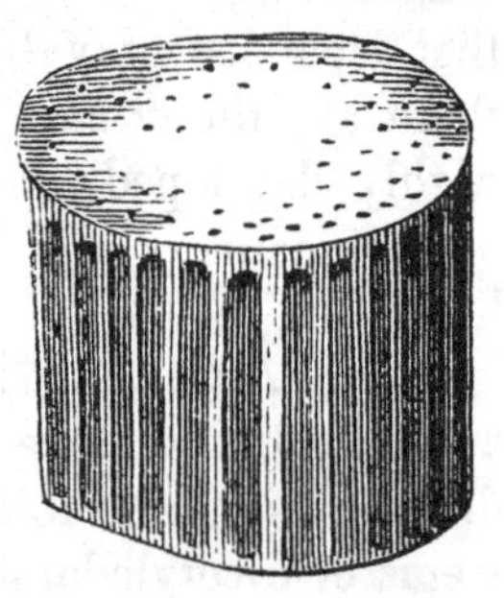
35

(49.) *Medullary Sheath.*—The fibres which compose the medullary sheath, appear to retain their vitality for a long time after the pith has been exhausted and become dead; and the tracheæ which abound in it may even be unrolled in old and dry wood.

(50.) *Wood.*—The woody layers seldom, if ever, contain perfect tracheæ; but they are composed principally of elongated cellular tissue, traversed by ducts of various kinds. As the tree becomes aged, the innermost layers grow darker and more solid, and are then termed the "Heart-wood," or "Duramen." The outer layers, which are called the "Alburnum," remain soft and pale, and are rejected by workmen as being unsuited to economic purposes. The variously coloured fancy woods employed by the turner consist of the heart only, the alburnum in the ebony, even, being quite white.

Each zone is principally composed of cellular tissue towards its inner, and of vascular tissue towards its outer parts: and each is supposed to be as a repetition of the parts formed during the first year's growth. In the common sumach (*Rhus typhinum*), especially, the cellular or inner part of each zone has precisely the same appearance as the pith, which is here of a peculiar brown colour and easily recognised. But as there are no tracheæ among the vessels in the outer part of the zones, whilst these are abundant in the medullary sheath, the analogy alluded to is not perfect.

Some woods contain scarcely any ducts, as many Coniferæ; and the delicate material of which rice-paper (as it is called) is composed, consists entirely of cellular tissue. This curious substance is procured from the herbaceous stems of a species of Æschynomene, growing in China. The whole stem is about an inch thick, and resembles a mass of pith covered by a very thin epidermis. There is, however, a central column of real pith running through it. By means of some sharp instrument, the stem is cut spirally round the axis into a thin lamina (*fig.* 36.), which is then unrolled, and may be made up into sheets containing about a foot square of surface.

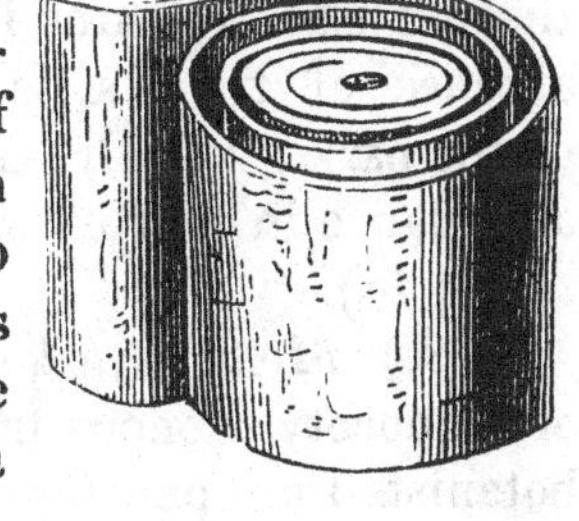

(51.) *Medullary Rays* (see *fig.* 24.). — These form what carpenters term the "silver grain" in wood, and are generally distinctly traceable in dicotyledonous trees. They may be seen passing in straight lines from the centre to the circumference, but cannot be traced continuously to any great extent in a vertical direction. They ap-

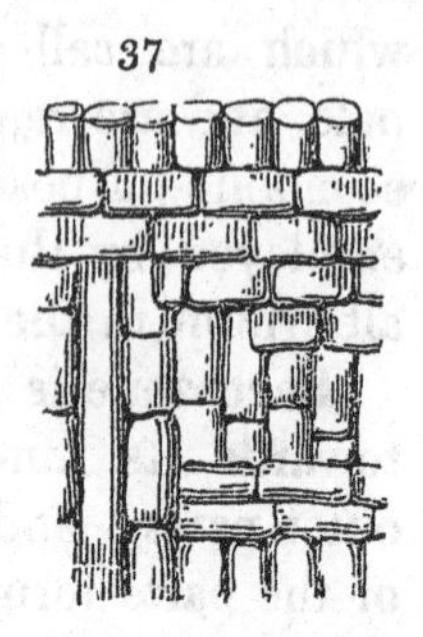
37

pear rather as isolated patches of cellular tissue, arranged in laminæ of one or more cells in thickness, placed at right angles to the concentric woody layers (*fig.* 37.). The cells are elongated in the direction of the rays. In some climbers, where the stem is twisted, the rays are curved from the centre to the circumference.

(52.) *Bark.*—The layers which compose the bark, are formed on a reverse plan to that of the woody layers, their outer portion being chiefly cellular, and their inner more vascular. The last formed or innermost, is termed the "Liber," the rest bear the general name of "Cortical layers." These layers are capable of greater or less distension, according to the nature of the tree; and in some cases the fibres are so far separated as to represent a sort of lace-work, as in the *Daphne lagetto.* In the lime tree, the inner layers, when separated by maceration, form the common bass, or matting, used by gardeners. The outer layers of the birch, beech, and other trees, are thrown off in thin membranous laminæ. In oaks, elms, and a multitude of others, the old bark remains in a rugged cracked state. The absence of tracheæ is a nearly universal characteristic of the bark; but Dr. Lindley has detected them in great abundance in that of the pitcher-plant (*Nepenthes distillatoria*).

(53.) *Monocotyledonous Stems.*—The complete want of monocotyledonous trees in our climate, has debarred botanists the opportunity of examining their structure so particularly as they have that of Dicotyledons; and, perhaps, even yet, the exact course of the woody fibres distributed through the trunk, is not accurately understood. It was supposed until lately, that the newest fibres were placed nearer the centre than the old ones, throughout the whole of their length (*fig.* 38. *a*); but M. Mohl has recently shown that this cannot be the case. He observes that the fibres cross each other before they pass into the leaves; and therefore supposes that the newest fibres are always nearer

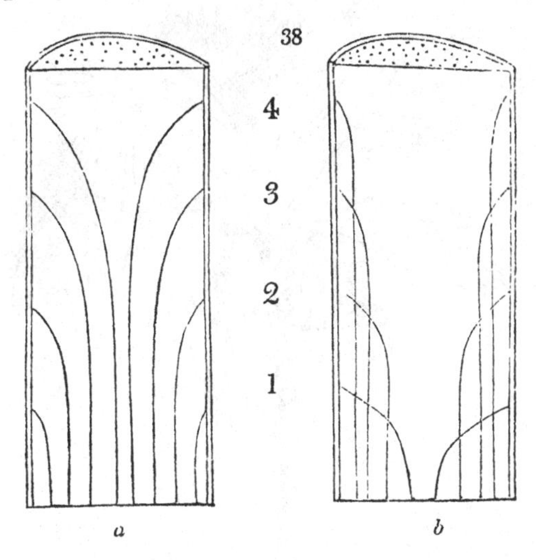

to the circumference than the old ones, at the bottom of the trunk, but that they cross them as they ascend, and then curve outwards and pass into the leaf (*b*).

Those monocotyledonous stems which have no branches, and are supplied with nutriment entirely from the leaves at the summit, continue of nearly equal thickness throughout their whole length, as in the lofty palms (*fig.* 39.), whose trunks are a long cylinder, crowned by a splendid mass of foliage. But those which are branched, become thicker below than above, as in dicotyledonous trees. The same may be said of such Monocotyledons as the grasses, whose stems are clothed with leaves throughout their whole length. It has, indeed, been generally asserted that the trunks of many monocotyledonous trees do not increase in thickness after they have risen above the surface of the soil; but such an assertion does not appear to have received a satisfactory confirmation. It is easier to believe that their increase is very slow, and that the fresh materials are always equally distributed from the top to the bottom—the diameter of the terminal bud increasing as the trunk lengthens. We find that even the trunks of old dicotyledonous trees, below the part where the boughs

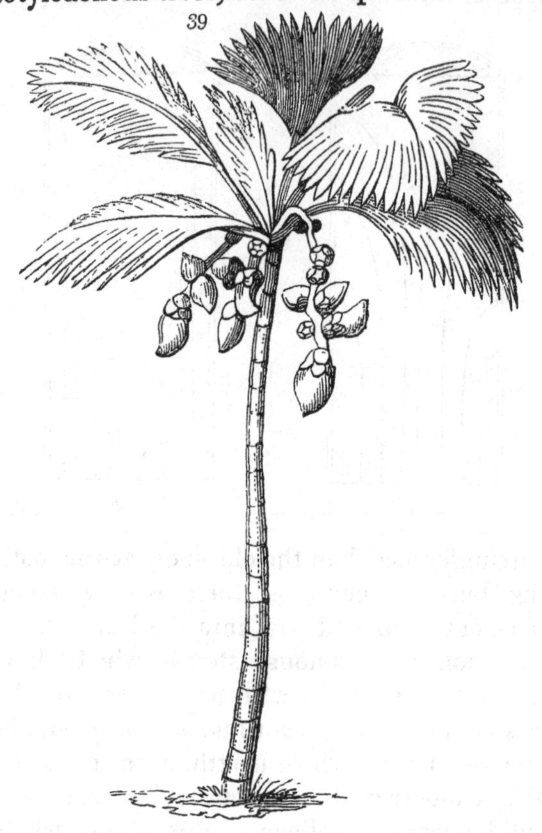

Lodoicea sechellarum.

set on, are nearly cylindrical, or frustra of very elongated cones, when compared with the portions above them. Mirbel has figured the trunk of a monocotyledonous tree which has become completely invested by a climber whose branches have grafted together into a reticulated cylindrical mass. This specimen has been considered to illustrate the fact, that the stem could not have increased at all in thickness after it had become so closely embraced. But something of the same sort may occasionally be observed even in dicotyledonous trunks, where they have become completely invested by ivy, whose branches intertwine and graft together, though perhaps not so completely as in the case of the creeper alluded to. That Monocotyledons increase very slowly in thickness may readily be conceived, but so do the trunks of dicotyledonous trees, after they have acquired a great age.

(54.) *Forms of Stems.* — The more usual character of dicotyledonous stems, is to taper off gradually from the base towards the summit, and they thus approximate to the form of a very lengthened cone. On the other hand, the stems of woody Monocotyledons, with few exceptions, approximate to the form of a cylinder. Some stems, however, in the early stages of their growth, and many herbaceous stems during the whole period of their duration, are variously angulated, and

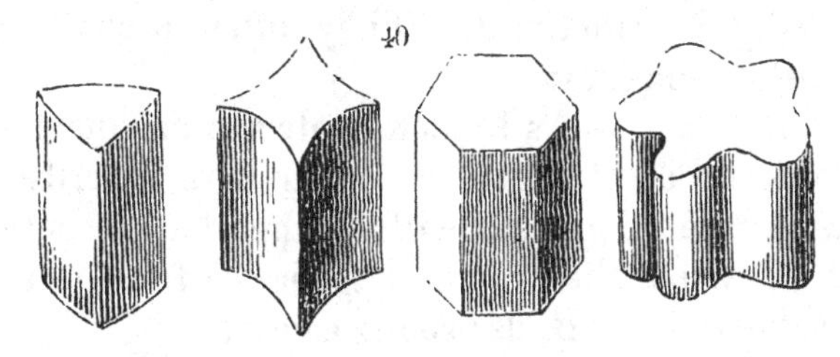

channelled (*fig.* 40.). This is frequently owing to some peculiarity in the development of the cellular tissue of which the bark is composed.

(55.) *Directions of Stems.* — The original tendency of aërial stems, is vertically upwards; but many are too weak to support themselves in that position, and, in consequence, either trail upon the ground, or cling to the surrounding herbage, by means of tendrils, hooks, and various other appendages; which are frequently modifications of the leaf. There are certain stems, also, which, by continually twisting in a spiral manner, twine themselves round the trunks and branches of neighbouring trees and shrubs, and are thus supported to a great height. The spiral which these stems describe, is termed

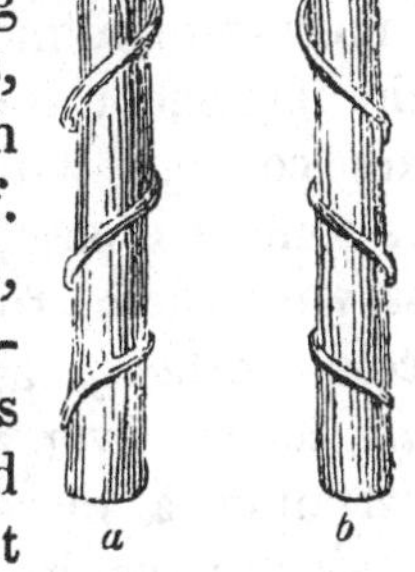

"right-handed" (*fig.* 41. *b*), or "left-handed" (*a*), according as its coils appear to *rise* from left to right, or from right to left, to a person supposed to be placed in its axis; or, if we were to hold the spiral in an upright position before us, then the coils of a right-handed spiral will seem to *descend* from the left towards the right, and those of a left-handed spiral to descend from the right towards the left.

(56.) *Knots, Internodia, and Joints.*— Many stems are swollen at intervals, where the leaves are attached, and such swellings are termed "knots." The space which intervenes between two knots, is an "internodium." "Joints" are also swollen parts, where the tissue is less firm than elsewhere (*see* art. 25.), and may easily be fractured. They often occur immediately below the knots.

(57.) *Buds.* — As branches always originate in the development of "buds," we shall here describe these bodies, before we proceed with further details concerning stems, of which the branches appear to form, as it were, mere subdivisions. Buds usually consist of several scales, or rudimentary leaves, closely wrapped round an axis; and within these are other leaves, in a still more rudimentary state, which are destined to assume a more highly developed condition than the outer scales of the bud. It is the outermost scales which thus serve to protect the innermost and more delicate parts, from the inclemencies of the weather. Some are covered with down, which may, as some suppose, be effective in preserving them from the intensity of cold; others, as the horse-chestnut, are coated over with gluten, which is certainly a more effectual protection against moisture; and perhaps this is the end which these scales best fulfil in most cases, as their closely im-

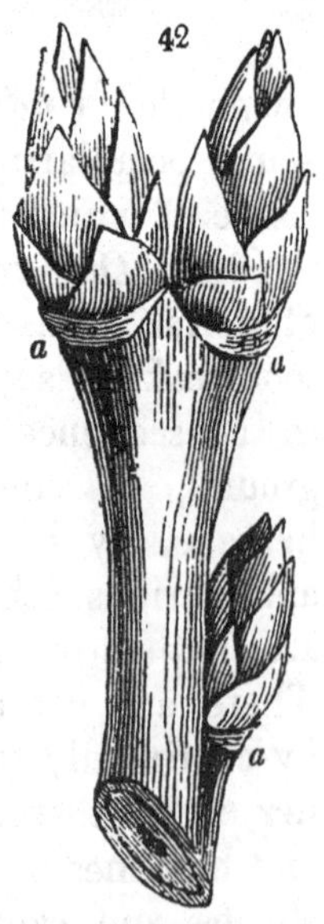

bricated condition, would seem to indicate. Buds are sort of nascent germ, originating within the stem, from the surface of which they ultimately protrude, and are developed (*fig.* 42.).

In ordinary cases, buds are formed about the places where the leaves unite with the stem; and they are most frequently situate immediately above the "axil" of the leaf—or place where this union occurs (*fig.* 42. *a*). In some plants, however, the buds are produced on the sides of the axils; and, in some, even within the space covered by the leaf-stalk, where, consequently, they lie concealed until the leaf falls. Buds may, however, be developed, under peculiar circumstances, from any part of the stem; and such are called "adventitious" buds, to distinguish them from those which are formed in the ordinary way.

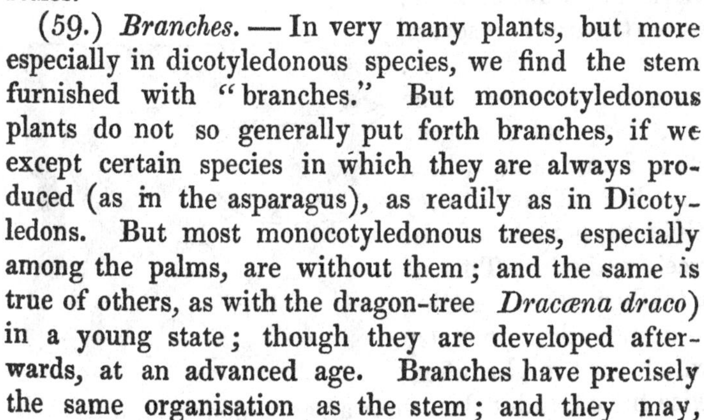

43

(58.) *Shoots.* — In the early stages of their development, branches are termed "shoots;" and, when they rise from underground stems, and their leaves assume the form of scales, as in the common asparagus (*fig.* 43.), the shoot is termed a "turio." In this plant, the leaves are never further developed; but buds are formed and branches proceed from the axils of the scales.

(59.) *Branches.* — In very many plants, but more especially in dicotyledonous species, we find the stem furnished with "branches." But monocotyledonous plants do not so generally put forth branches, if we except certain species in which they are always produced (as in the asparagus), as readily as in Dicotyledons. But most monocotyledonous trees, especially among the palms, are without them; and the same is true of others, as with the dragon-tree *Dracæna draco*) in a young state; though they are developed afterwards, at an advanced age. Branches have precisely the same organisation as the stem; and they may, in fact, be considered as so many partial stems engrafted into the main trunk. Originating, as we have stated, from buds, their disposition round the stem must depend upon the arrangement of the leaves, to which we shall allude when we treat of those organs. We may, however, remark, that branches are never so symmetrically arranged as leaves; because a great many buds are never developed at all. This arises from the unfavourable circumstances under which many are placed, for receiving a sufficiency of air, of moisture, and more especially of light. The consequence is, that those which originate on the lower parts of the stem, are either much stunted, or become abortive.

(60.) *Development of Branches.*—When a branch is not developed, where a bud has been formed, the latter still continues to live; and, in dicotyledonous trees, is carried outward with the increasing bulk of the stem, and awaits at the surface for a proper opportunity, when a sufficient quantity of light, or of some other requisite, may enable it to "break" into a branch. This fact is familiar to every horticulturist, and is the foundation of the principle upon which he regulates the pruning of his trees. If a section of the stem be made at the point where an undeveloped bud is seen to protrude, it will show the course which the bud has followed in passing from the centre outwards, marked by a line or wake, which traverses the several layers (*fig.* 44.). Hence, branches of the same age, may have originated from buds which have been formed at very different periods of the tree's growth. This is a further cause, tending to destroy the symmetry which they might otherwise have exhibited in their arrangement round the axis

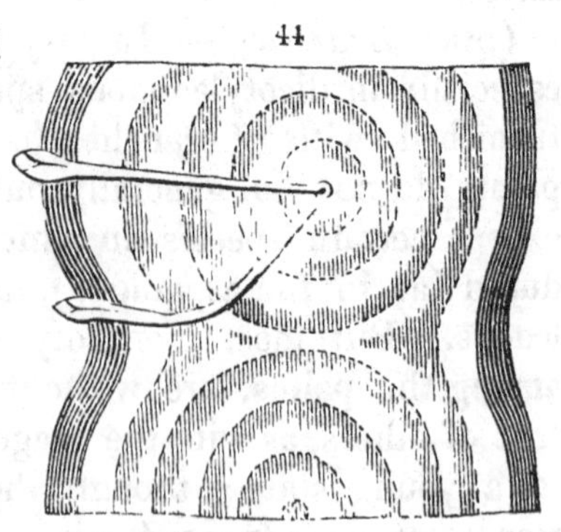
44

of the stem. In the annexed diagram (*fig.* 45.), *a* represents a bud, developed on a branch which is one year old; and this branch is seated on another which is two years old, and which originated from a bud of the same age as *b*, which has not yet been developed.

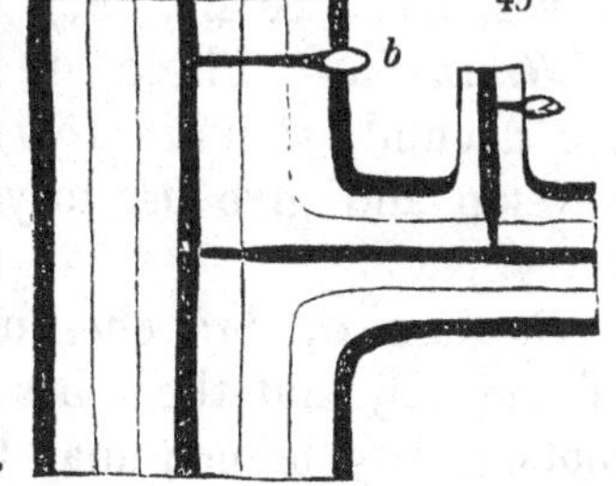

(61.) *Direction of branches.* — The general contour given to the whole foliage of trees,—which bears the name of "cyma," depends upon the angle which the branches make with the stem at their point of union, combined with the degree of rigidity which they possess. When they stand out at various angles, more or less approaching to a right angle, they are termed "divergent;" and, when such branches are rigid, a rounded form is given to the cyma, as in the oak and elm. When the angle is more obtuse, they are said to be "patent," or "spreading." If they rise at a very acute angle, and are packed close together into the pyramidal forms assumed by the cypress and Lombardy poplar, they are called "appressed." When they are very long, and so flexible as to bend by their own weight, they are "pendant," as in the weeping willow. But in that variety of the common ash, which is also called "weeping," the branches are rigid, and possess a natural tendency downwards, from their very origin, and are in this case termed "depressed."

(62.) *Modifications of Branches.*—

Thorns. — When a bud is imperfectly developed, it sometimes becomes a short branch, very hard and sharp at the extremity, and is then called a "thorn." We must not, however, confound the "prickle" with the thorn. The former of these is a mere prolongation of cellular tissue, from the bark, and may be considered as a compound kind of pubescence (art. 31.); whilst the thorn, containing both wood and bark, is an organ of the same description as the branch itself. "Spines" originate in the transformation of leaves, &c. (see art. 78.).

Runners.— These are branches which trail along the ground, striking root at intervals, where the buds develop and give rise to young plants, as in the strawberry.

Suckers are branches originating below the surface of the soil, and their base in consequence soon emits roots. Any branch may be made to assume this character artificially, by confining a portion of it below the surface; as the horticulturist is aware when he forms his "layers."

(63.) *Subterranean Stems and Branches.*—There are some stems and branches, which, instead of rising upwards, continue under ground, and creep horizontally below the surface of the soil. These are very generally mistaken for roots, and are usually termed "creeping roots;" but they may readily be distinguished from roots, if not by their internal structure, at least by their external appendages. They are mostly furnished with scaly processes, or other traces of a degenerated and modified form of the leaves; and they produce buds, and often throw up branches which rise above ground; or else they themselves ultimately take a tendency upwards, and become true aërial stems; a good example of which occurs in the common reed (*Phragmites communis, fig.* 46.). The swollen rhizomata of this plant runs among the turf of our fens, forming large tubes through the masses cut for burning. They are furnished at intervals with pale membranous scales. or rudimentary leaves; and fibrous roots are given off from all the knots. So soon as the rhizoma takes an upward tendency, it contracts its dimensions, and ultimately rises above ground as a slender stem, invested with long green leaves. The term "rhizoma or root-stalk," is equally applied to prostrate stems, as in the iris tribe, and in some ferns, where the upper surface gives rise to the leaves, and the lower to the roots; and also to the completely subterraneous stems which throw up stalks and leaves at intervals

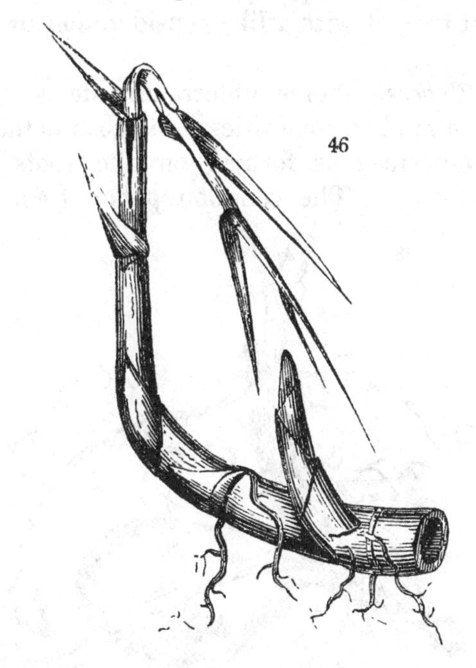

(*fig.* 47.), as in the *Carex arenaria*, *Elymus arenarius*, &c., — plants of inestimable utility in certain regions,

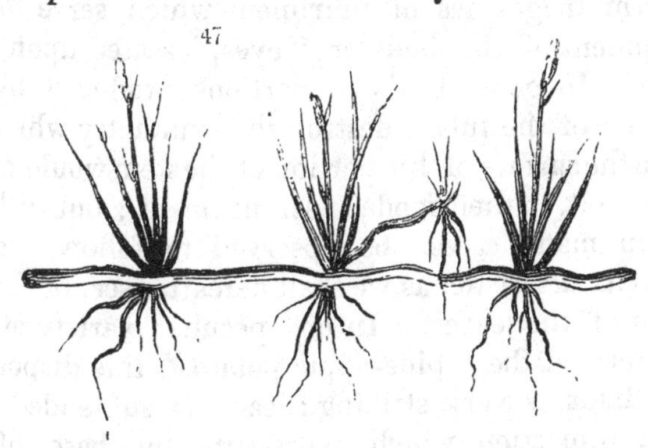

where they serve to bind the shifting sands of the sea shore, which would otherwise drift before the wind, and form irruptions over the neighbouring land. The common but noxious couch-grass is another familiar example of the kind, equally interesting to the botanist, though not treated with a like consideration by the agriculturist.

(64.) *Tubers.*—Some subterranean stems or branches terminate in swollen nodosities, analogous to those which we have described as formed on the roots of some plants (art. 40.). The common potato (*fig.* 48.) is a

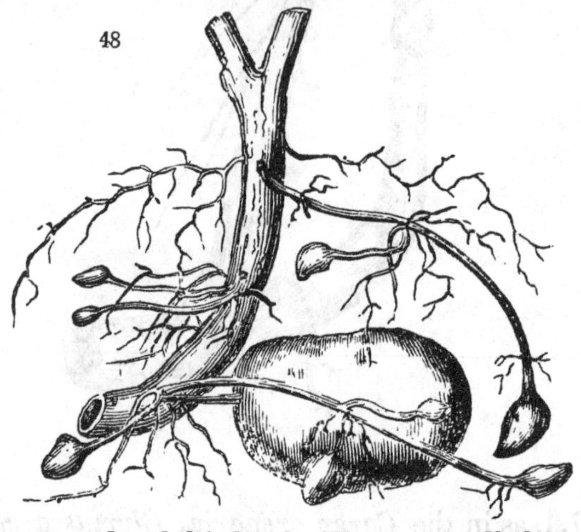
48

familiar example of this kind. These are called "tubers," and form magazines of nutriment which serve for the development of the buds or "eyes," seated upon their surface. In general, the distortions produced by the formation of the tuber, destroy the symmetry which the buds on the surface of this portion of the stem would otherwise exhibit, in their mode of arrangement; but still they may, in many cases, be observed to follow a spiral course, characteristic, as we shall hereafter see, of the disposition of the leaves. In one peculiar variety of this tuber, termed the "pine-apple potato," this disposition of the buds is very striking; each is subtended by a swollen projection which represents the base of the leaf-stalk, in whose axil we may consider it to have been formed. In turnips, radishes, &c., this tuberous development originates in the lowest portions of their stems, which are placed either wholly or partially below ground; whilst in the Kohl-rabbi (a variety of cabbage), the effect is produced on a part of the stem which is entirely above ground.

(65.) *Bulbs.* — The buds of some plants are subject to a peculiar modification. Instead of expanding into branches and leaves, in the usual way, the rudimentary parts of which they consist, become depositaries of nutriment, — swelling preternaturally, but still continuing in a condensed or undeveloped form. In this state they are termed "bulbs;" and are sometimes found on the stems, and in the axils of the leaves, as in the Orange-lily (*Lilium bulbiferum*); and even among heads of flowers, as in a variety of the common onion. The bulbs, however, with which we are most familiar, as of lilies, hyacinths, onions, &c., contain the whole of the ascending organs of these plants in a condensed form, with their roots proceeding from a flat disk below (*fig.* 49.). The chief differences among

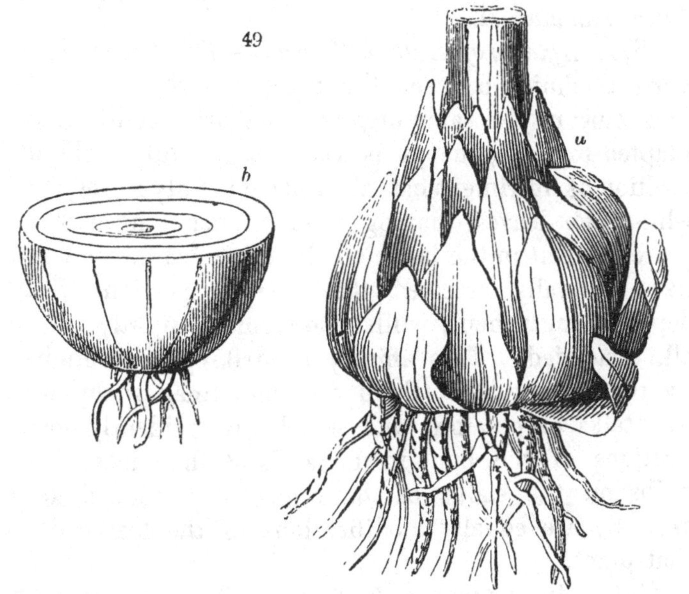

bulbs depend upon the rudimentary leaves of which they are composed, being either in the form of succulent or fleshy scales (*a*), as in the lily; or in concentric coats (*b*) which completely surround the axis, as in the onion; in the latter case, also, some of the outermost laminæ are thin and membranous. The young bulbs, or "cloves," as gardeners term them, are produced, as we should expect, by the development of fresh buds in the axils of the scales or laminæ of the old bulb.

(66.) *Cormus.* — The name of "cormus," is given to the swollen base of some stems of monocotyledonous plants, or rather to the condensed state of the whole stem (*fig.* 50.); which is developed underground, and assumes the general appearance of a coated bulb, as in Crocus and Colchicum, where it is sometimes erroneously termed a "solid bulb;" or else it resembles a tuber, as in the common *Arum maculatum.*

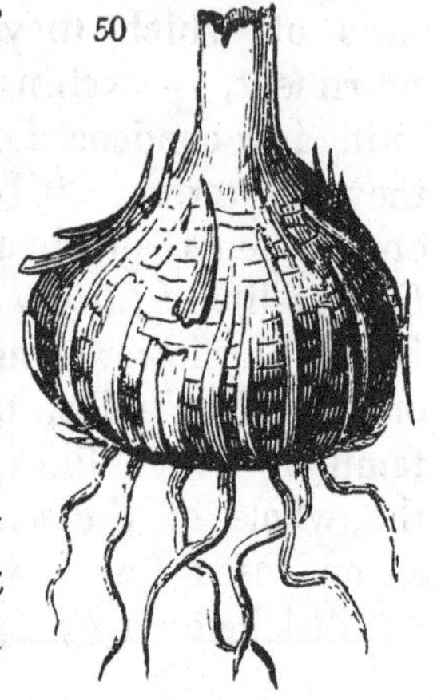

(67.) *Affinity of Bulb to Tuber.* — There is evidently a great affinity between the tuber and the bulb; each consisting of the same organs, peculiarly modified, and adapted to analogous purposes. In the tuber, the deposition of nutriment has taken place mainly in the stem, whilst the leaves, having received none, have disappeared. But in the bulb, on the other hand, the leaves have generally received the greatest portion of the deposited nutriment, whilst the stem is slightly, or not all, distended. This affinity is strikingly exemplified by the little tubers which are sometimes produced on the stalks of potatoes, and which are evidently modifications of the buds in the axils of their leaves; the bulbs on the stalks of the orange-lily alluded to in art. 64., are equally modifications of the leaf-buds of that plant.

(68.) *Appendages to the Stem.* — The various organs which we have just been describing, ought rather to be considered as "modifications," of certain parts of the stem, than as distinct appendages to it: but we have now to mention a long list of organs, situate on some part or other of its surface, which are properly styled "appendages" to the stem or ascending axis. Diversified as these organs are in their forms, and even in their functions, they may all be considered as modifications or transformations of one fundamental organ, of very general, though not universal occurrence, viz. the leaf. In order to obtain a general notion of the varied appearances assumed by this organ, we must suppose that some of the materials which compose the stem have become detached from the rest, and are then given off at the surface, in the form of distinct organs.

CHAP. III.

NUTRITIVE ORGANS — *continued.*

LEAVES, SIMPLE AND COMPOUND (69.). — VERNATION (71.). — FORMS OF LEAVES (74.). — PHYLLODIA (75.). — TRANSFORMATION OF LEAVES (78.). — VENATION (81.). — DISPOSITION AND ADHESION (82.). — NUTRITIVE ORGANS OF CRYPTOGAMIC PLANTS (84.).

(69.) *Leaves.* — In by far the greater number of plants, these organs consist of thin flattened expansions, in which the vascular portion, termed "veins," or "nerves," is arranged in a kind of network, having the interstices filled up with cellular tissue — here termed the "parenchyma;" and the whole is invested with the epidermis. In Dicotyledons, the vessels proceed immediately from the medullary sheath. In a few rare examples, as in the *Dracontium pertusum* (*fig.* 51.), the parenchyma imperfectly fills up the interstices between the veins, and large holes are left through the leaf (*a*). In the most curious and interesting *Hydrogeton fenestralis* (*fig.* 52.), an aquatic of Madagascar, the parenchyma is so little developed, that the leaf appears to consist entirely of the veins, and resembles those skeletons of leaves which are sometimes prepared by maceration in water. A large proportion of trees produce fresh leaves in the spring, which afterwards fall in the autumn; such are termed "deciduous," in contradistinction to "evergreens," which are never entirely divested of leaves. No plant, however, retains its leaves for more than two or three years; but as the leaf-fall in evergreens is partial, consisting perhaps of one half or one third at a time, there are always a sufficient number left on the tree, to keep it clothed with perpetual verdure.

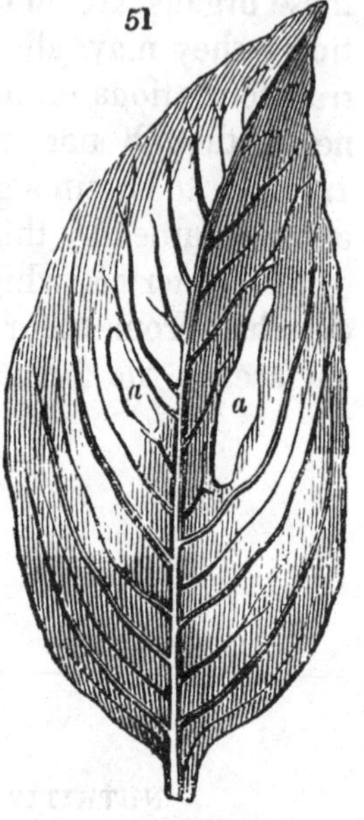

In succulent plants, the vessels which quit the stem to form the leaf, diverge in different planes, and the leaves in consequence consist of solid fleshy masses of cylindrical and other solid forms, instead of flattened laminæ.

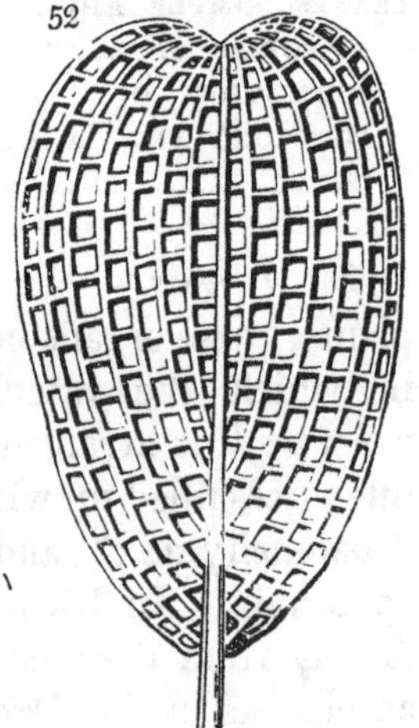

The complete leaf consists of two parts: the leaf-stalk, or "petiole;" and the expansion, or "limb." There is often an alteration in the colour and texture of the petiole at the point where it is attached to the branch, and sometimes a slightly swollen protuberance. This is termed an "articulation;" and it is at that part that a disunion takes place at the period of leaf-fall, and a "scar" is left upon the stem. But where no articulation exists, the withered petiole remains a long time attached to the stem before it falls off and leaves the scar. Some petioles are termed "clasping," when they are attached for some extent around the stem; and they form "sheaths," when they wholly embrace it, as in the grasses. In some, a membranous limb-like expansion occurs on each side of the petiole, which is then said to be "winged." The limb in general is similarly constructed on each side of the midrib; but to this there are striking exceptions, as in the leaves of Begonia (*fig.* 53.), Epimedium, &c.

53

(70.) *Simple and compound Leaves.* — The most obvious classification of leaves, is into "simple" and "compound." The limb of the former consists of one piece only (*fig.* 54.), which may either be entire at the margin (*a*), or variously indented (*b*); and attached to the stem with or without the intervention of a petiole: in the latter case it is said to be "sessile." Compound leaves (*fig.* 55.) are made up of one or more pieces, called "leaflets," each of which is *articulated* to the petiole; and the degree to which it is compounded, depends upon the number of times in which the main petiole branches, before the leaflets are attached to its ramifications. Hence we have the simply (*a*), doubly, triply (*b*), &c. compound leaf.

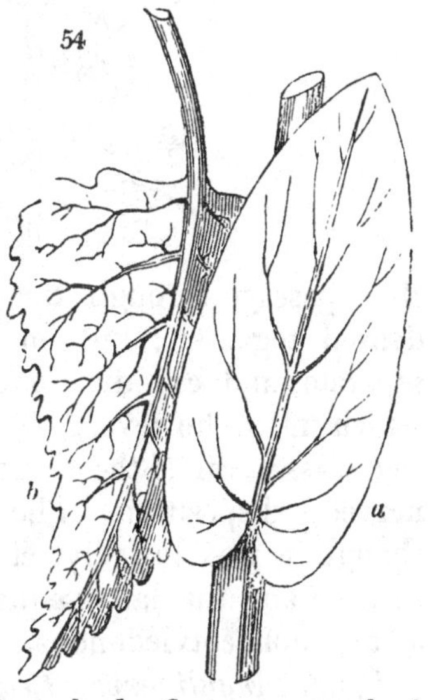

(71.) *Venation or Nervation of Leaves.*—The distribution of the vascular tissue through the limb of the leaf is termed its "venation" or "nervation"—the course of the vessels bearing some resemblance to the distribution of veins and nerves in certain parts of the animal structure. The bundles of vessels which compose the veins, maintain a nearly parallel course in

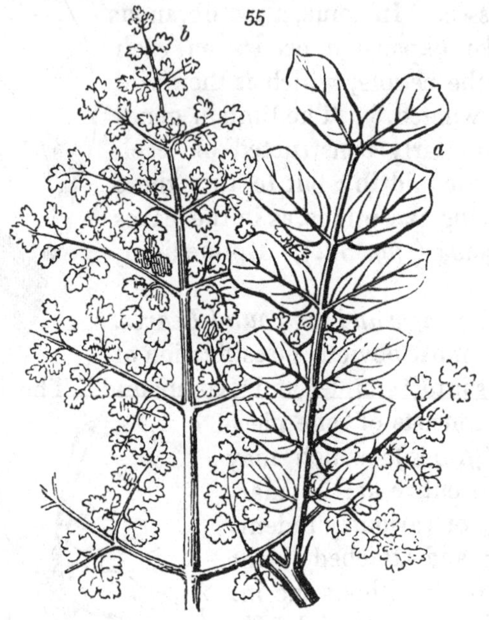

their passage through the petiole, and are closely condensed together; but on arriving at the limb, they separate, and are distributed in various ways; all of which may, however, be referred to one or other of two classes, called the "angulinerved," and the "curvinerved," disposition. The former of these is eminently, though not exclusively, characteristic of dicotyledonous plants; and the latter equally predominant among such as are monocotyledonous.

(72.) *Angulinerved Leaves.*—In these, the vessels, after entering the limb, either branch off immediately from the apex of the petiole, and form several strong veins; or they form one midrib, from which secondary veins are given off on either side, and which at their origin, maintain a straight course for a short distance, however they may afterwards be curved (*fig.* 54.). The angle at which they diverge is generally acute, towards the apex of the limb, and their mode of ramification bears a resemblance to the branching of trees. This kind of nervation may be subdivided into four subordinate groups, which are important in regulating the conditions upon which some of the principal forms of leaves depend.

(a.) *Penninerved.*—Here the midrib is continued to the extremity of the limb, and the primary nerves branch off from it on either side, throughout its whole length (*fig.* 56.). The breadth of the leaf is chiefly regulated by the size of the angle at which the nerves quit the midrib, being narrower in proportion as this angle is more acute. The contour of the limb is also defined by the proportion which the different nerves bear to each other on quitting different parts of the midrib. This form of nervation is by far the most usual, and regulates the structure of many compound leaves. In these the main petiole may be likened to the midrib of a simple leaf, with its parenchyma only partially developed round the secondary nerves, so that it becomes split up into separate leaflets. Compound leaves are pinnate, bi-, tri-, &c. pinnate, according to the degree of subdivision to which the branching of the petiole extends. But when the limb of a leaf is merely subdivided, without being completely separated into distinct leaflets, the terms applied to designate the degree of subdivision are "pinnatifid," "bi-, tri-, &c. pinnatifid." In pinnate leaves, the leaflets are frequently arranged in pairs, on opposite sides of the petiole, with or without a terminal leaflet.

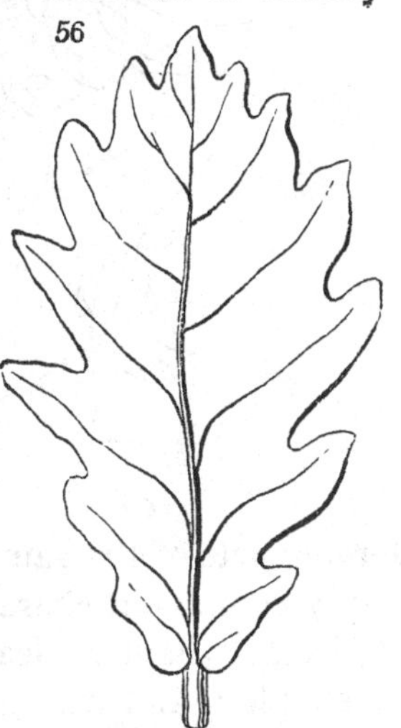
56

The intimate relation which subsists between simple and compound leaves, is well exemplified in some cases, where two or more contiguous leaflets become grafted together, and thus reduce the usual extent of the subdivision to a lower degree. This may be often seen in some species of Gleditsia (*fig.* 57.), where dif-

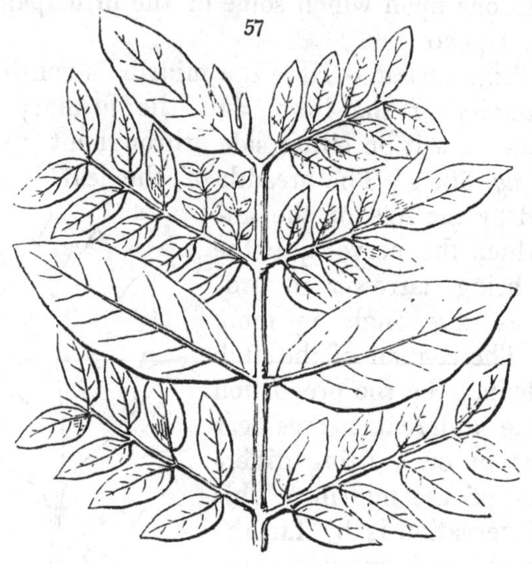

ferent parts of the same leaf assume a simply, doubly, or triply compound character. It is difficult in some cases to decide whether a leaf should be considered compound, or simple; and it is usual to account all leaflets which are articulated to the petiole, as parts of a compound leaf, even though they may be reduced to one in number, as in the case of the orange; but those which are not articulated, even though they may be otherwise distinctly formed, are considered as subdivisions only of a simple leaf. Where these articulations exist, each leaflet falls separately from the main petiole, when this also becomes detached from the stem; but where the leaflets are not articulated to the petiole, the limb falls entire, with the petiole attached.

(b.) *Palminerved.*—Instead of forming a midrib, the

vessels here diverge from the extremity of the petiole into several (usually three or five) equally strong nerves, which are afterwards subdivided in a penninerved manner (*fig.* 58.). The whole system of venation here resembles that of a compound penninerved leaf, whose leaflets have become grafted together into one limb. This nervation stamps the character of the palmate leaves.

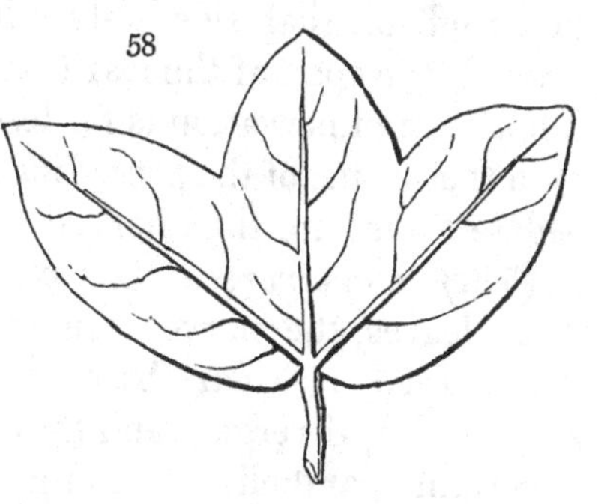

(c.) *Peltinerved.*—The vessels in this case diverge in a plane which is inclined to the direction of the petiole; and in proportion as the angle of inclination approaches a right angle, the limb of the leaf is more symmetrically formed, round the point where the petiole is attached to it (*fig.* 59.). Where the angle is acute, the nerves which diverge on the side nearest to the petiole are the shortest, and the limb is proportionably contracted. From this nervation originate the peltate leaves.

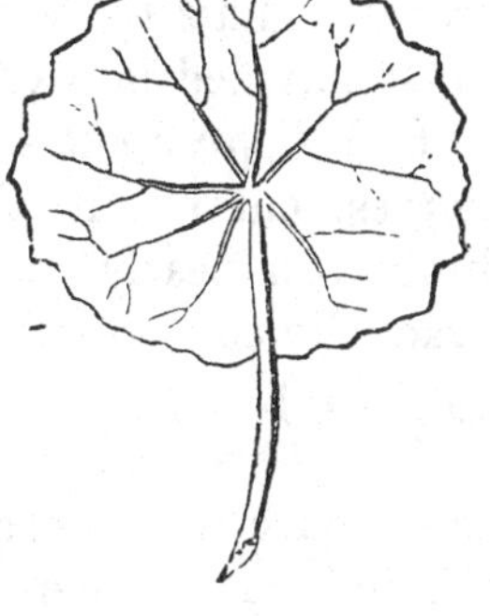

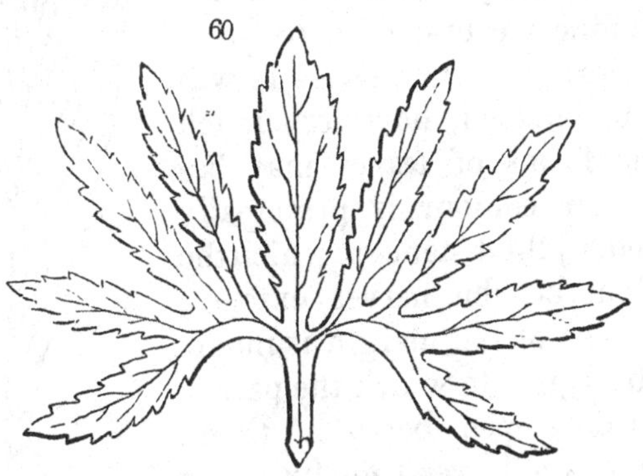

(d.) *Pedalinerved.*—In this case there is no decided

midrib, but the vessels diverge in two strong lateral nerves, from which branches are given off, on that side only which is towards the apex of the leaf (*fig.* 60.). This form of nervation is far less common than either of the preceding. The pedate leaves are thus nerved.

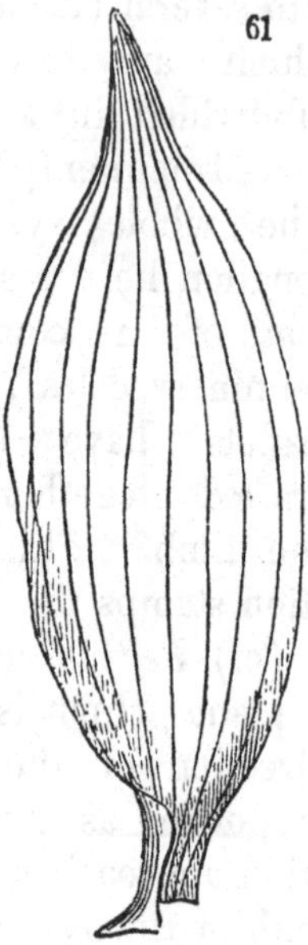

61

(73.) *Curvinerved Leaves.* — In these leaves, the nerves are more or less curved at their base, or point whence they diverge; and they retain a certain parallelism among themselves, as well as a simplicity of structure, which very readily distinguishes them from the angulinerved leaves. This mode of nervation may be subdivided into two classes.

(a.) *Convergent.* — Where several nerves, curved at the base of the limb, run nearly parallel to its margins, and proceed gradually converging towards its apex (*fig.* 61.).

62

(b.) *Divergent.* — Where the vessels collectively form a midrib to the limb, and numerous simple nerves diverge from it in a pinnate manner, but maintain nearly a parallel, or somewhat curvilinear course (*fig.* 62.).

(74.) *Forms of Leaves.* — It will easily be understood, how very much varied the forms of leaves may become. Their contour is principally determined, by the distance to which the ramifications of the nerves extend; and the shape of the margin is modified, by the degree to which the parenchyma is developed between them. Thus, in ovate leaves (*fig.* 63.), the margin of *a*, which is only slightly indented, is said to be "toothed;" that of *b*, which has the indentations deeper, is called "divided," or "incised;" and *c* is termed "partite." Where the limb is almost severed into separate segments, each portion, when tolerably large, is also termed a "lobe," and the angle at which the lobes meet is the "sinus." When the teeth are large and regular, they are termed "serratures;" and when these are rounded, "crenations." Thus, a vast number of terms, most of them of very simple construction, and easy comprehension, are used, for expressing a variety of different modifications, by which these and other organs of planst may be accurately defined.

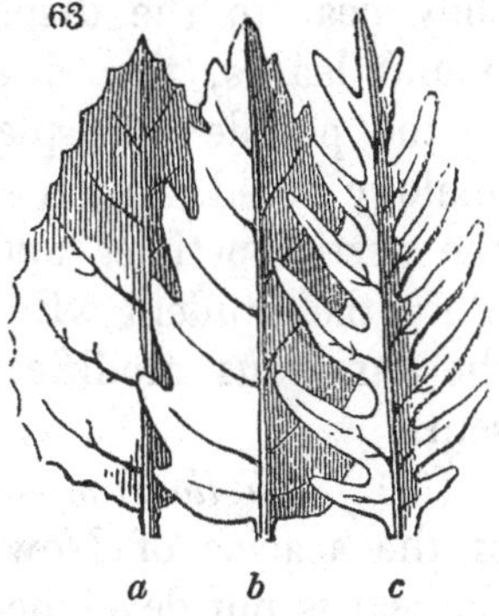

The leaves on different parts of the same plant often differ in shape; and even those on the same part are sometimes subject to great modifications, according as they are influenced by the peculiar circumstances under which they are developed. Thus, we may occasionally find three varieties, among the radical leaves on the same plant of horse-radish (*Cochlearia armoracia*), where the marginal indentations vary as much as in *fig.* 63. In general, however, the leaves of the same plants, or at least on the same parts of a plant, retain a sufficient constancy in their character, to enable us to use them for the purpose of discriminating between species which are very closely allied. It would not be in character with our present undertaking, to enter more minutely into any description of the forms of leaves; but we recommend all who wish to pursue this subject further, and to become acquainted with those technicalities of the science which are necessary for the purposes of accurate description and descrimination of species, to notice the dependence which the forms of leaves possess upon the conditions of their venation. In the first place, they should remark the general contour of the limb, without reference to its marginal incisions; then they should consider the character of the incisions, and the relation they bear to the disposition of the veins. In compound leaves, the degree to which the subdivisions of the petiole take place must be considered, and the analogy noted, which exists between the disposition of the partial petioles and the venation of simple leaves. Thus the student will soon learn to fix in his memory the numerous modifications of form which leaves present.

(75.) *Phyllodium.*—There are some plants, as many of the acacias of New Holland, in which the limb of the leaf is not developed, but the petioles themselves are laterally compressed; and so much flattened out as to assume the appearance of a limb; except that they affect a vertical position instead of a horizontal one, and that there is no apparent difference between their two surfaces in colour, or other characters. In young plants

64

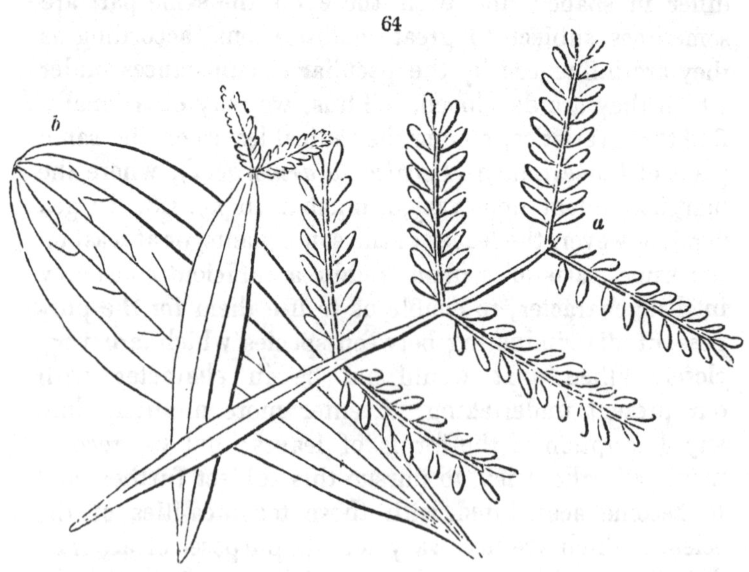

of this description, however, and occasionally also in old ones which have been freely pruned, we may observe all the intermediate states or varieties between a doubly compound leaf (*fig.* 64. *a*) and the simply expanded petiole just described (*b*); the latter being more dilated in proportion as the leaflets of the limb are fewer in number. These flattened petioles are termed "phyllodia," and the character of their venation, corresponds very closely with that of the curvinerved leaves of monocotyledonous plants. The non-development of the limb is also common in some species of Monocotyledons, which are nevertheless, capable of producing one. The *Sagittaria sagittifolia* (*fig.* 65.), an aquatic of this class, has the limb developed at the summit of those leaves only, which reach above the surface of the water, all the rest consisting merely, of strap-shaped expansions of the petioles.

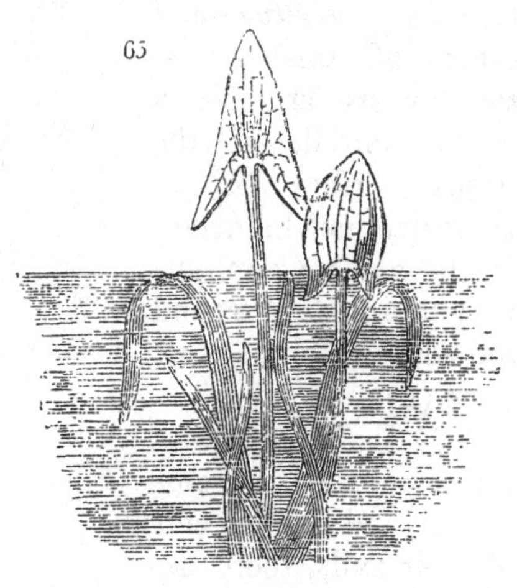

De Candolle considers the greater number of sheathing leaves, which are not furnished with distinct limbs, to be only petioles; and although such are found in several Dicotyledons, as in *Ranunculus gramineus*, *Lathyrus nissolia*, the whole genus Bupleurum, and some others, yet they are more especially characteristic of Monocotyledons, where he supposes the development of a true limb to the leaf to be comparatively rare; though it certainly occurs in the Arum tribes, Sagittariæ, and some others. Some limbless petioles are cylindrical and pointed like the leaves of a rush.

(76.) *Foliaceous Branches.*—The phyllodium is not the only substitute which nature provides, to supply the absence of a perfect leaf. In some plants, the leaf is completely abortive, and becomes a small dry scale, incapable of performing any of the proper functions of this organ. In these cases, the branches themselves become flattened, and assume the appearance of leaves (*fig*. 66.). In the common butchers'-broom (*Ruscus aculeatus*), and others of this genus, the flowers are seated in the middle of the upper surface (*a*) of these flattened branches. In the genus Xylophylla they are placed round the edges of similar organs (*b*).

(77.) *Stipules.* — At the base of some leaves, and on each side of their axils, there are appendages of a foliaceous character, sometimes resembling the leaflets of compound leaves, and sometimes like small membranous scales (*fig*. 67. *a a*). These are

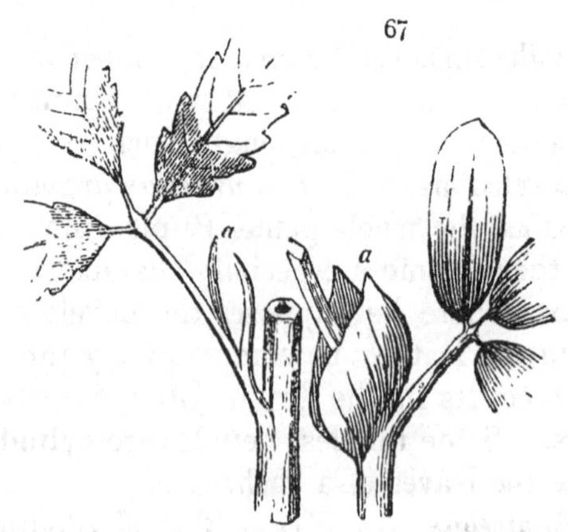

termed "stipules," and are very characteristic of certain groups of plants, but are entirely wanting in others. They are never found on any Monocotyledons, or on any dicotyledonous plant where the petioles are "sheathing."

(78.) *Spines.* — Some leaves, which do not freely develop in the usual manner, assume a dry hardened appearance, and pass into spines, as in the common furze; just as some abortive branches have been stated to assume the character of thorns (art 62.). In the berberry (*fig.* 68.) all the intermediate states (*s*) be-

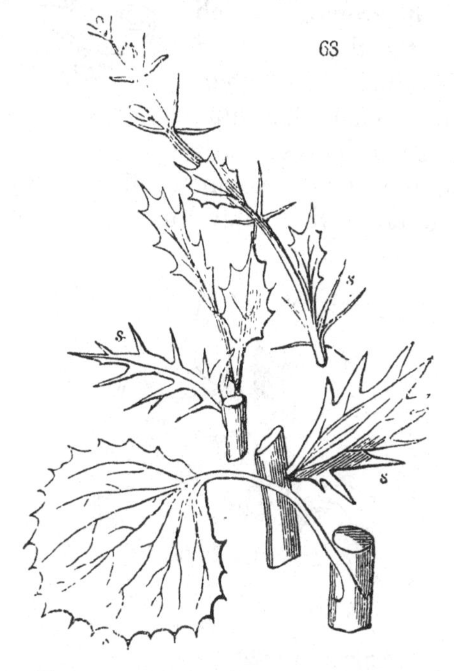

tween a well-developed leaf and the hard spine, may be distinctly traced, on vigorous suckers of a year's growth.

(79.) *Tendril.* — In some leaves, the midrib is protruded beyond the apex of the limb, in the form of a filamentous chord, and, in many cases, the limb entirely disappears, and the whole petiole is transformed into what is termed a "tendril." These organs serve to support the weak stems of certain plants, by twisting round the branches of others, in their neighbourhood.

In the *Lathyrus aphaca* (*fig.* 69.) all the leaves become tendrils, except the first pair in the young plants, which are compound, and have two or three pairs of leaflets. Occasionally an odd leaflet (*b*) is developed on the tendrils, in a later stage of growth, which further indicates the origin of the organ on which it is seated. A provision is made for supplying the want of leaves in this plant, by an unusual development of the stipules (*a*), which are so large that they might readily be mis-

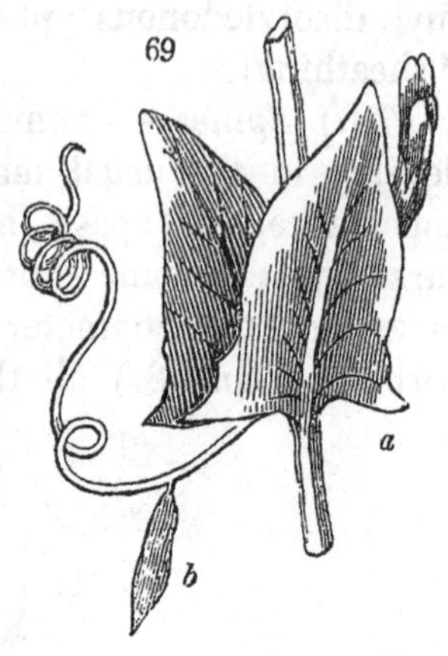

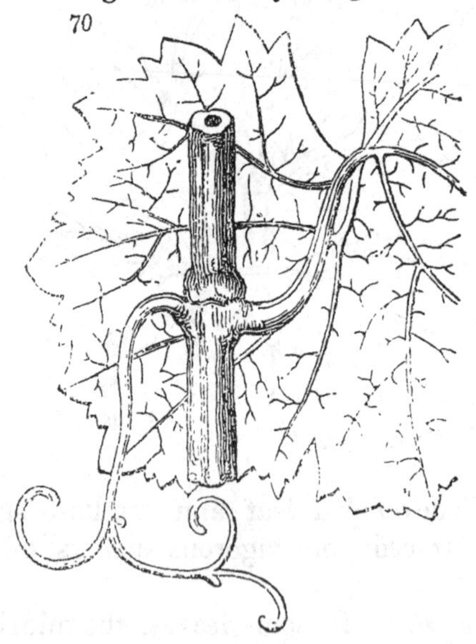

taken for real leaves. All tendrils, however, do not originate in the modification of the leaf; but some are derived from an altered condition of the stipules, as in the cucumber; others, from a transformation of the branches or peduncles, as in the vine (*fig.* 70.). In fact, they may result from any of the caulinar appendages, which become lengthened out at their extremities into filiform flexible cords, more or less spirally twisted.

(80.) *Pitcher.* — Of all the metamorphoses which the leaf is found to undergo, the singular productions called "pitchers" are the most curious. The annexed

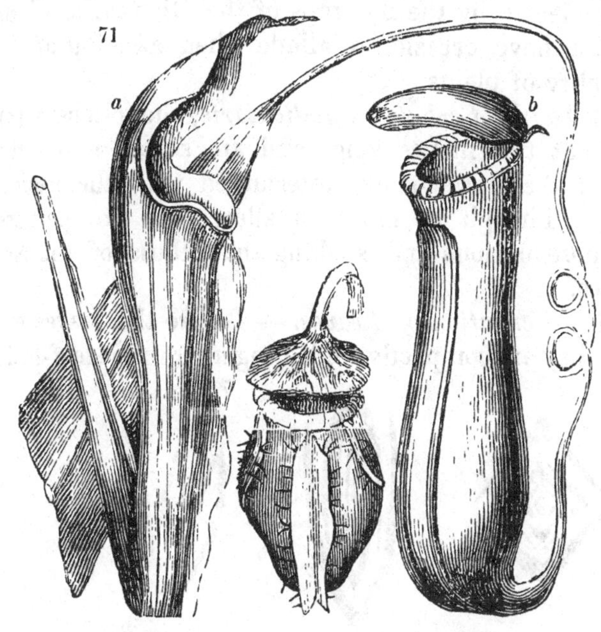

cut (*fig.* 71.) represents three different forms of these organs.

(*a.*) In the genus Sarracenia, nearly the whole leaf resembles a funnel, with the upper extremity crowned by a membranous expansion, tapering to a point.

(*b.*) In the Nepenthes, or true pitcher-plant, the pitcher (*b*) is placed at the extremity of a tendril, terminating a winged petiole. It is crowned with a membranous lid, which is closely shut in the early stages of its growth, but is afterwards raised, and does not again close the aperture. These pitchers, in some species, are six or seven inches in length, and have the lower portion of the inner surface, of a glandular structure, which is constantly secreting a subacid liquid. In this liquid a number of insects are continually drowned;

and, strange as the idea may seem, it has been conjectured, that the providing of such animal manure for the plant, is one object which these singular appendages were intended to accomplish. There is, certainly, a striking analogy between this result, and the still less equivocal object effected by the fly-traps of the Dionæa, to which we shall have occasion to allude when speaking of the irritability of plants.

(*c.*) In the *Cephalotus follicularis*, the pitchers (*c*) are about two inches long, and are seated round the base of the flower-stalk, intermixed with the radical leaves. Though so much smaller, they are perhaps still more curious and striking than those of the Nepenthes.

(81.) *Vernation of Leaves.* — Before the leaves expand, they are compactly folded together in the leaf-bud;

and the various modes in which this takes place, is called their "vernation." The folds or plaits either lie in a longitudinal direction, parallel to the midrib; or they are transverse, so as to bring the apex and base towards each other. Different terms are applied to the various modes of vernation, some of which, however, are seldom employed in descriptive botany. The appearances represented in the annexed cut (*fig.* 72.) are among the most striking and important, and are obtained by making a transverse section through the leaf-buds of different plants: *a,* plicate; *b,* equitant; *c,* imbricate; *d,* involute; *e,* revolute; *f,* obvolute; *g,* circinate.

(82.) *Disposition of Leaves.* — Although the term "radical leaves," is applied to those which are seated close to the ground, and appear to spring from the summit of the root itself, yet all leaves do, in fact, originate upon the stem or branches. In a general way we may refer their disposition to one or other of two modes: either "verticillate," when more than one is attached to the stem at the same altitude, or about the same horizontal plane; or "alternate," when they are so dispersed upon the stem that no two are seated precisely in the same horizontal plane. When the number of leaves in the same plane does not exceed two, and these lie on contrary sides of the stem, they are said to be "opposite." Leaves are frequently so arranged, one above another, as to form two or more ranks down the stem; and sometimes they appear to follow the direction of spiral lines which coil round it. These

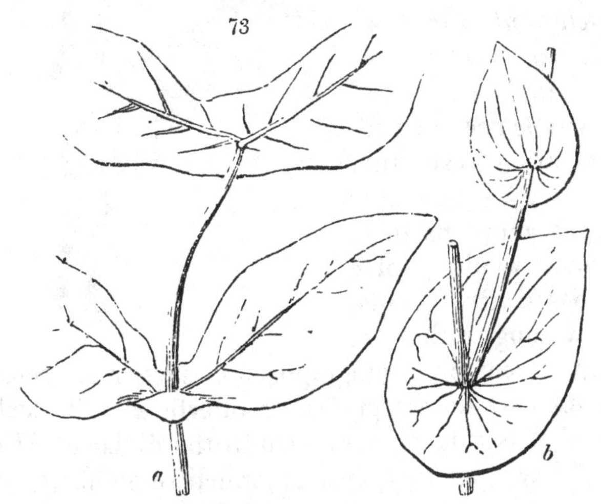

different appearances receive appropriate names in descriptive botany, which it does not fall in with our plan to dilate upon; but, before we have concluded this part of our subject, we shall enter somewhat more fully into the details of a theory, which has been proposed for reducing under general laws, all the modes which are observable in the distribution of foliaceous appendages.

(83.) *Adhesion of Leaves.* — In some species where the leaves are opposite, we find them "connate," or grafted together by their bases (*fig.* 73. *a*), so as completely to surround the stem; and in other species, where they are alternate, and without a petiole (sessile), the edges at the base of the limb extend round the stem (*b*), and are united together. Both these cases are termed "perfoliate;" the stem seeming as it were to penetrate the leaf. In some plants, the middle of the leaf adheres to the stem, through a greater or less extent, whilst its edges are free (*fig.* 74.). The leaf is here said to be "decurrent," and the stem "winged."

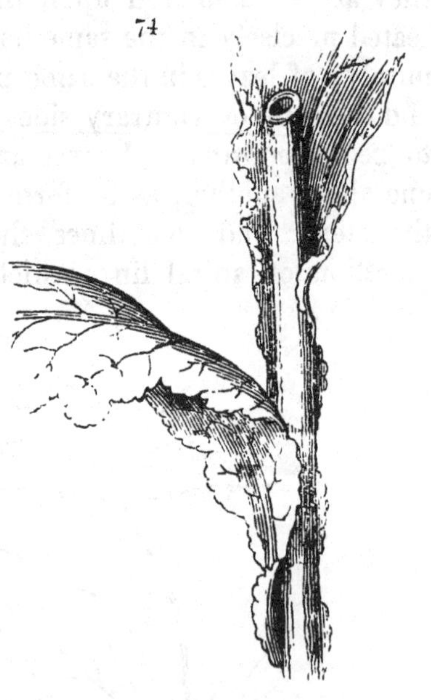

(84.) *Nutritive Organs of Cryptogamic Plants.*— In art. 36. we have already stated nearly all that it will be necessary for us to mention respecting the organs of cryptogamic plants; a more particular account would involve us in descriptive details, which belong rather to the department of phytography and systematic botany, with which we do not profess to interfere. The higher tribes of these plants, contained in the division "Ductulosæ," have green expansions, much resembling leaves in their general appearance, and like them possessing stomata; but differing from them very considerably in some respects, especially in bearing the fructification upon their surface. These have therefore received a distinct appellation, and are called "Fronds;" and that part of a frond which is analogous to the petiole, is

75

termed the "Stipes." In some cases, as in the tree ferns of tropical climates (*fig.* 75.), the bases of the decayed fronds form a tall trunk, which is termed their "caudex;" but when this portion creeps upon the ground, as in the humbler forms of our own climate, it has received the name of "rhizoma." In several tribes the fronds possess nerves, but in many cases they are composed entirely of cellular tissue. The vernation of the fronds of most ferns is peculiar, and termed "circinate" (*fig.* 72. *g*). It consists in having all the extremities of its different subdivisions, as well as the whole frond itself, rolled inwards. The lower tribes of cryptogamic plants, included in the division "Cellulares," are very homogeneous in their structure, and of different degrees of consistency — from highly gelatinous, to tough and leathery. When they consist of a plane membranous lamina, as in the Lichens, this is termed a "thallus" (*fig.* 76.); but when more

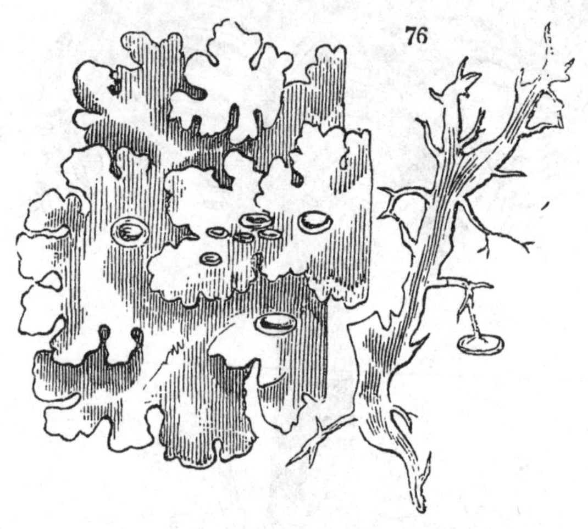

or less branched, the name of frond is retained. They are either terrestrial, aquatic, or marine. Many of them are parasitic, seldom green, and without stomata.

CHAP. IV.

REPRODUCTIVE ORGANS.

FLOWER BUDS (85.). — INFLORESCENCE — MODES OF (86.). — FLORAL WHORLS—PERIANTH (92.). — GLUMACEOUS FLOWERS (96.). — STAMENS AND PISTILS (97.). — DISK (101.). — FLORAL MODIFICATIONS (102.). — ÆSTIVATION (104.).

(85.) *Flower Buds.*—NUMEROUS examples are perpetually occurring, in which the attentive observer of nature may catch a glimpse of the mysterious connection which subsists between the organs of nutrition and reproduction, in plants. Instances continually present themselves, of flowers whose separate portions are singularly characterised, by possessing an intermediate condition, partly leaf-like, and partly like those variously coloured appendages which constitute the blossom. By an accurate examination of these and other "monstrosities," as all deviations from the ordinary conditions of vegetation are termed, it has been clearly ascertained, that the organs of reproduction and nutrition are merely modifications of some one common germ, which may be developed according to circumstances, either in the form of a flower-bud, or of a leaf-bud. In the latter case we have shown, how this body becomes a branch and leaves; and we have now to explain the conditions and characters of those several organs which are developed from the flower-bud, and collectively termed the "inflorescence." It would be equally erroneous for us to call the flower-bud a metamorphosed state of the leaf-bud, as to say the leaf-bud was an altered condition of the flower-bud; and we are nearer the truth, when we consider each of them to be a peculiar modification of the same kind of germ, adapted in the one case to perform the functions of nutrition, and in the other, those of reproduction.

Flower-buds ought consequently to make their appearance on similar parts of the stem and branches with the leaf-buds, viz. in the axils of the leaves; and the development of each will present us with analogous phenomena. However different in their external characters, still the various parts of the inflorescence must bear a strong affinity to those of the foliaceous appendages on the branch.

(86.) *Inflorescence.*—In this term we include, not merely the flower which proceeds from the development of the flower-bud, but also the stalk on which it is placed, and any of those other various appendages upon it, which are always more or less distinct from true leaves. The more general term for the flower-stalk is "peduncle," but the term "pedicel" is also used in a restricted sense, where there are partial flower-stalks seated upon a common peduncle. The flower-stalk is more or less dilated at the apex, when there are several flowers closely crowded upon it, and without distinct pedicels, as in the order Compositæ. Such dilatations of the flower-stalks receive the general name of "receptacles," but other terms are specially applied to some of their modifications. The foliaceous appendages on the peduncle, which more or less resemble the stem-leaves, but which are also sometimes reduced to the condition of mere scales, are called "bracteæ." The flower terminates the pedicel, and is composed of certain foliaceous appendages, which are still further removed from the character and condition of leaves, than the bracteæ. The analogy which exists between the various parts of a leaf-branch and those organs which compose the inflorescence, is very often exhibited in certain monstrosities of the rose; where we find the central parts of the flower, instead of assuming their usual character, become developed as a branch. It sometimes happens that this monstrous development will again make an effort to pass to the state of a flower, and then the central parts will a second time assume the condition of a branch. In the Water-avens (*Geum rivale, fig.* 77.) this description of monstrosity is particularly frequent; and, indeed, it may be often seen in many other flowers.

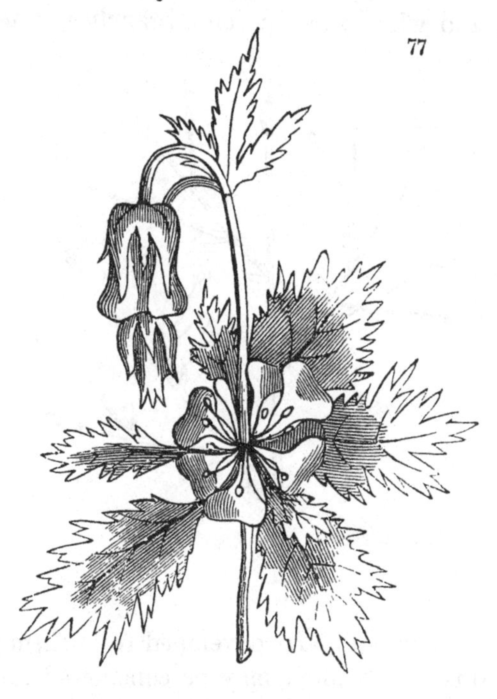

77

(87.) *Modes of Inflorescence.*—From what we have said, it will be evident that the term inflorescence, is either applied to the appearance presented by the general disposition of all the flowers on a plant taken collectively, or it is confined to certain groups of flowers which are found on different branches; or, lastly, it is restricted to solitary flowers produced from separate buds. In order to understand the general law, which regulates the distribution of flowers under every form of inflorescence, according to the vague application of this term in descriptive botany, it will be well to consider the manner in which we may conceive it possible, for a succession of buds to become developed upon the main stem, or any of the branches. Assuming any bud (*fig.* 78.) from which the stem or given branch is developed, to be the "primary" bud (No. 1.) of the series we are investigating, then "secondary" buds (Nos. 2.) are developed from the axils of the leaves or bracteæ; and when these become branches, "tertiary"

78

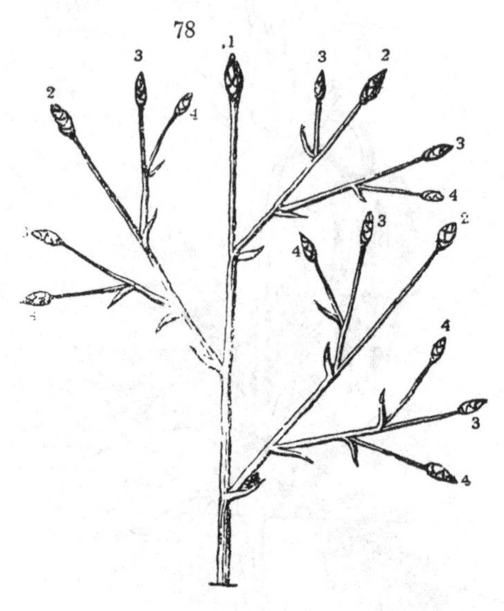

buds (Nos. 3.) are similarly developed from them; and so on. In this way a plant may be considered capable of indefinitely multiplying the number of its branches, and also of extending them to any length, by the continued development of the terminal bud at the extremity of each of them. Trees continue to develop a succession of buds in this manner for many years together, without producing flower-buds; but some trees, and all herbaceous plants, soon produce flower-buds, and then the branches on which they occur are abruptly terminated. Now, it appears to be a general rule, that when the buds of one order cease to develop as branches, by becoming flower-buds, then the buds of the next order, which are developed round the axis of the former, likewise terminate in flower-buds. Thus, if No. 1., after developing a branch and leaves, ultimately becomes a flower-bud, then every bud (Nos. 2, 3, 4, &c.) which terminates the branches developed round its axis, will also ultimately terminate in flowers. Now, in the common definition or notion of Inflorescence, we either include only a certain aggregation of branches, all of which terminate in flowers, or else we include one or more of those branches, whose terminal buds still continue to develop as leaf-buds, without ever becoming flower-buds. It has been supposed, indeed, that there are two distinct modes of inflorescence, in one of which the terminal bud does, and in the other it does not, become a flower. But this depends merely upon the vague manner in which we include under our definitions of inflorescence, a greater or less number of buds of different orders of development. If we admit a bud which does not terminate in a flower, to be the primary bud included in the inflorescence, then we have what has been termed the "Indefinite inflorescence," because the main axis continues to develop indefinitely, whilst the lateral buds alone terminate in flowers. But if the main axis, of what we choose to include within the inflorescence, terminate in a flower, then the "Terminal inflorescence" is the result. There are numerous modifications of both these kinds of inflorescence, which either depend upon the disposition of the leaves or bracteæ, in whose axils the flower-buds originate, or else upon the partial abortion, or peculiar development, of some or of all the secondary, tertiary &c. buds; and also upon other circumstances.

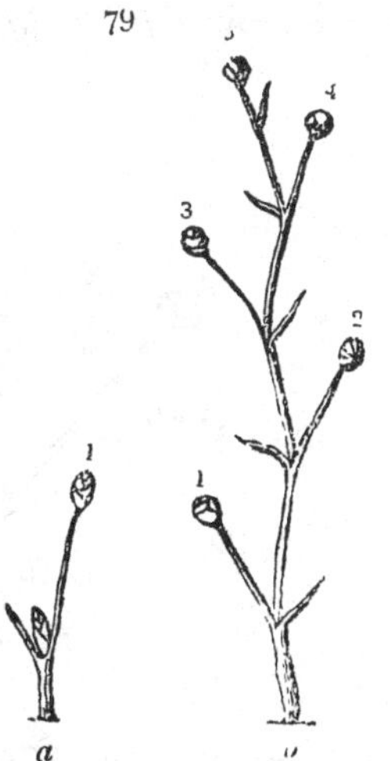

(88.) *The Terminal Inflorescence.* — The principal axis included in this inflorescence, terminates in a flower-bud, and the secondary buds are developed in the axil of each leaf or bractea, situated at the base of that portion of the branch which becomes a peduncle, and must therefore be placed immediately between a leaf and a flower (*fig.* 79.). If the secondary bud is not developed, the inflorescence must consist of a solitary flower (*a*). If the leaves are placed alternately on the axis, the peduncle of the flower will bear a single bractea at its base. If the secondary bud is developed (*b* No. 2.), it will terminate in a flower with a bractea at the bottom of its peduncle, bearing a tertiary bud in its axil; and this (No. 3.) may develop like the former; and so on. In this case, all the flowers will appear to stand opposite the leaves or bracteæ, upon a stem which seems to develop indefinitely; but which is, in reality, composed of a succession of branches or peduncles, originating from different orders of buds. Since No. 1. is the real termination of the main axis, and Nos. 2, 3, &c. are further and further removed from it, the order in which the flowers expand is from the centre outwards, and this has in consequence been termed the "Centrifugal inflorescence."

When the leaves or bracteæ are opposite or verticillate, in the terminal inflorescence, this is called a "cyme." When each secondary bud is developed from the axils of a pair of opposite bracteæ, and the tertiary buds originate in the same manner, and so on, the cyme is styled "dichotomous" (*fig.* 80. *a*). If there be a whorl of three

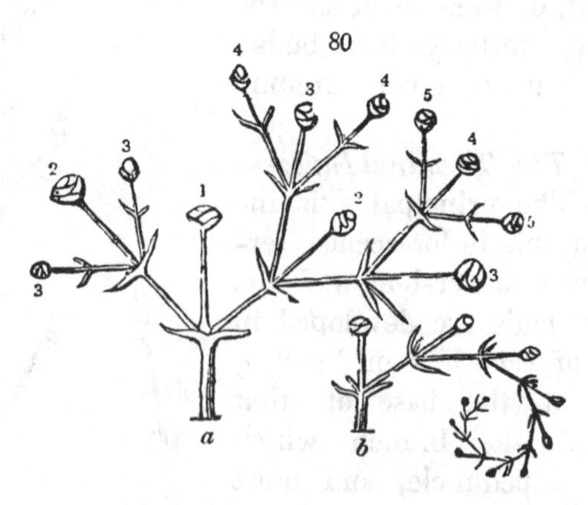

bracteæ, the cyme is "trichotomous," &c. If, however, one bud only is developed in the dichotomous cyme, and always on the same side of the axis, it assumes a peculiar character, termed "scorpioidal" (*b*).

(89.) *Indefinite Inflorescence.* — Here the terminal bud, of the main axis included in the inflorescence, continues to develop as a leaf-bud, until sooner or later it is exhausted, and the branch stops; but it does not pass to the condition of a flower-bud. If we first consider the case where the leaves are alternate, then the secondary buds in the axils of the leaves or bracteæ may either become flowers immediately (*fig.* 81. *a*); or they may be

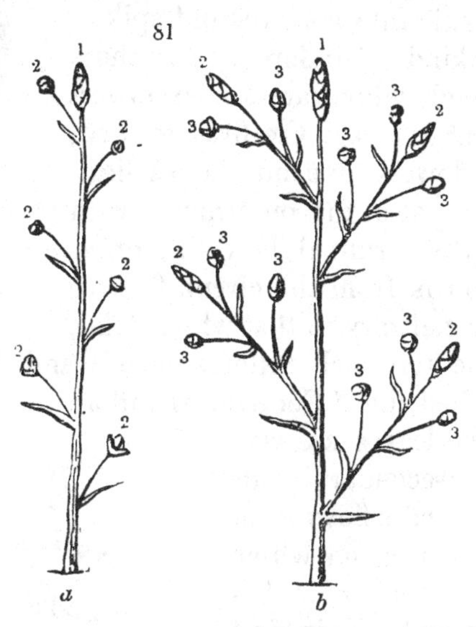

partially developed as branches (*b*) which give rise to tertiary buds; and these may become flowers, or branch in the same way as the secondary buds. When the secondary buds become flowers, without previously branching (*a*), the inflorescence is termed a "raceme," or "cluster," provided each flower has a pedicel; but it is called a "spike," if the flowers are sessile, or without pedicels. Where the secondary buds become branches, bearing flowers produced from tertiary buds, the raceme is called "compound" (*b*). A few of the subordinate varieties of these forms may here be noticed. In such plants as the willow, hazel (*fig.* 82.), and oak, the peculiar spike in which the flowers are arranged is termed a "catkin." In the tribe to which the common arum belongs (*Aroideæ*), the fleshy mass which forms the axis round which the flowers are aggregated in a spike, is termed the "spadix" (*fig.* 88. *b*). The small spikes in which the flowers of grasses are aggregated, are termed "spikelets" (*fig.* 95. *c*); and these, again, are arranged round a common axis into a compound spike.

82

In this kind of inflorescence, those secondary buds which are seated lowest on the main axis are the first formed, and their flowers expand the earliest. As these are also the outermost, with respect to the terminal bud, the order of expansion is from the circumference inwards, or contrary to that which takes place in the terminal inflorescence; and hence this has been called the "Centripetal inflorescence."

When the leaves are verticillate, the secondary buds may either become flowers, or produce branches, on which buds of a lower order become flower-buds. This kind of inflorescence is generally called "whorled," and is either simple or compound (*fig.* 83.).

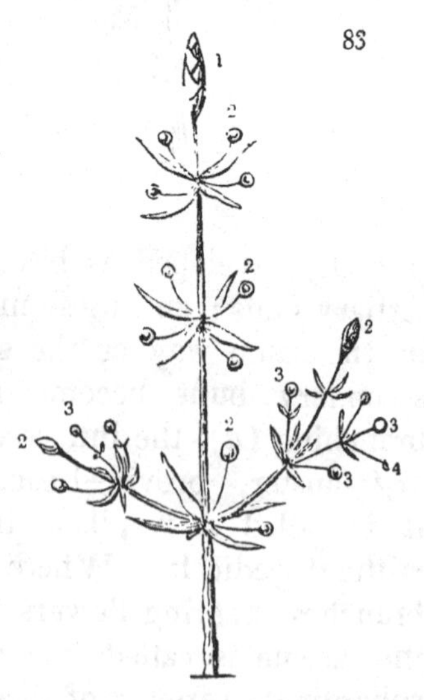

(90.) *Modifications of Inflorescence.* — It will be seen from what has been said, that the application of the term "inflorescence," is as indefinite as the use of the word "organ," which we equally employ, to signify the several parts of a plant, as well as the subordinate portions of which those parts themselves are composed. And thus, in some cases, we term a single flower the inflorescence; in others, an aggregation of flowers; or even include some buds which produce no flowers. Perhaps we might find terms, which would express more definitely the different orders of buds, included in our notion of inflorescence: and then, the flowers of all terminal inflorescences would be subordinate to buds of the first order; whilst the flowers of those which are styled indefinite, would commence only from buds of a second, third, &c. order. Each kind of inflorescence might be considered as simple, or as doubly, triply, &c. compound, according as one or more orders of buds were developed in the form of flowers. It might happen, that a terminal inflorescence, in which several orders of buds were developed (as *fig.* 79.), would contain fewer flowers than an indefinite inflorescence, in which one order only (as *fig.* 81. *a*) was developed. Both kinds also include several forms, strikingly similar in their general appearance, and which, in descriptive botany, have received the same names. Of these forms we may enumerate the following:

"Panicle." — When the secondary, tertiary, &c. buds are developed on long peduncles and pedicels, so that the flowers are loosely aggregated, or, as it were, scattered round the axis (*fig.* 84.).

84

"Corymb." —When the main axis soon terminates, and the secondary, tertiary, &c. buds form peduncles of such lengths, that the flowers which terminate them stand at nearly the same level. The peduncles are, of course, of different lengths, those towards the summit being the shortest (*fig.* 85.).

"Umbel." — When the main axis is so contracted

between the bracteæ, that all the secondary buds are crowded together, and developed from one point at its

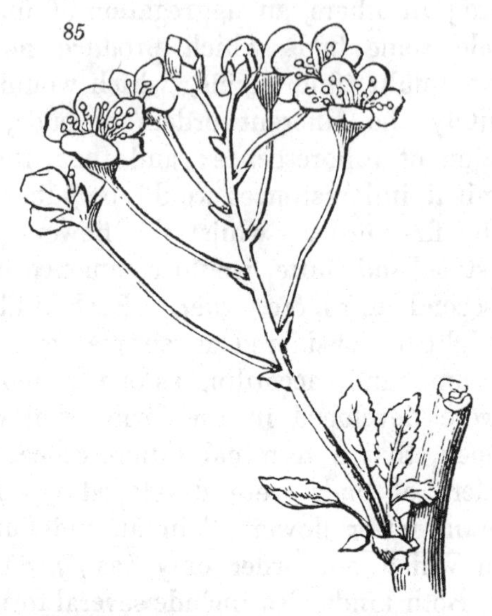

summit (*fig.* 86.). The pedicels are of the same length, so that all the flowers stand at the same level, as in the last case. When several small, or "partial"

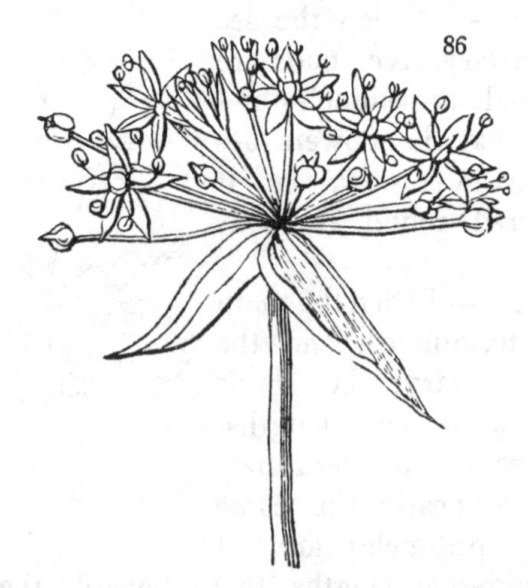

umbels, are themselves arranged in an umbelliferous manner round a common axis, the inflorescence is called a "compound Umbel."

An umbellate form, may evidently result also from a terminal inflorescence, where the leaves are whorled, and the secondary buds become flowers without producing tertiary buds. It often happens (as in the genus Euphorbia) that the main axis is crowned by an umbel of this description, whilst the lower part possesses the character of a raceme.

"Capitulum." — This form bears much the same relation to an umbel, that the spike does to the raceme; the pedicels of the single flowers being wanting, or scarcely distinguishable. The flowers are, in consequence, crowded into a dense head (*fig.* 87.).

(91.) *Bractea.* — We have said, as the flower-bud expands, a succession of various kinds of appendages, which depart more or less from the leafy structure, are developed round the peduncle, and that all of these would have become true leaves, if the bud had been impressed with the character of the leaf-bud. Of these appendages, the "bracteæ," as we stated (art. 86.), exhibit the closest approximation to the leaf itself, and, in many cases, are only nominally distinguishable from it, by their position alone. In general, however, they are of much smaller dimensions than the leaves, and are often reduced to mere scales. Sometimes they approach the appearances presented by the parts which compose the flower, and are brilliantly coloured. In the "cone" (*fig.* 137.), which is a modified form of the spike, having the flowers very closely arranged together, the bracteæ become large scales. These, in the fir tribe are coriaceous, and membranaceous in the hop.

When the bracteæ are arranged in a distinct whorl

round the peduncle, it is termed an "involucrum;" and in some cases they cohere by their edges, and thus form a single piece. Where the bractea, or rather involucrum, is very large, and completely envelopes the flowers, as in the Aroideæ, it is called a "spathe" (*fig.* 88. *a*). In the extensive order Compositæ, the little florets are crowded on a highly dilated receptacle, as in the common daisy and dandelion; and they are closely surrounded by an involucrum (*fig.* 87. *a*), composed of many bracteæ, which are either free, or adhere together, and the whole head has the appearance of a single flower. The cup in which the acorn is placed, is an involucrum, composed of several whorls of bracteæ, all adhering, and blended together into a solid mass (*fig.* 118.).

88

(92.) *Floral Whorls.* — The foliaceous appendages which succeed the bracteæ in the order of development, are brought close together, by the non-extension of the axis, so as to crown the summit of the flower-stalk with a series of whorls, partaking still less of the leafy character than the bracteæ (art. 86.). These whorls constitute the flower; and the portion of the axis on which they are seated, is termed the torus, which bears the same relation to a single flower, as the receptacle does to a head of flowers.

In flowers which possess the greatest number of whorls, such as those of the natural order Ranunculaceæ, we may distinguish four different kinds of organs; two of which, composing the outermost whorls, are collectively termed the "perianth;" and these are not essential to the fertility of the plant; but the two kinds which make up the innermost whorls, are absolutely requisite to secure the perfection of the seed. It is not necessary, indeed, that both the latter kinds should be found in the same flower, or even in different flowers seated on the same individual plant; but unless both exist, and can be subjected to a mutual influence, the fertility of the seed is never secured. A more accurate notion of these several whorls may be obtained, if we now examine the blossoms of a common ranunculus in greater detail (*fig.* 89. *a*). Here, the outermost whorl of the

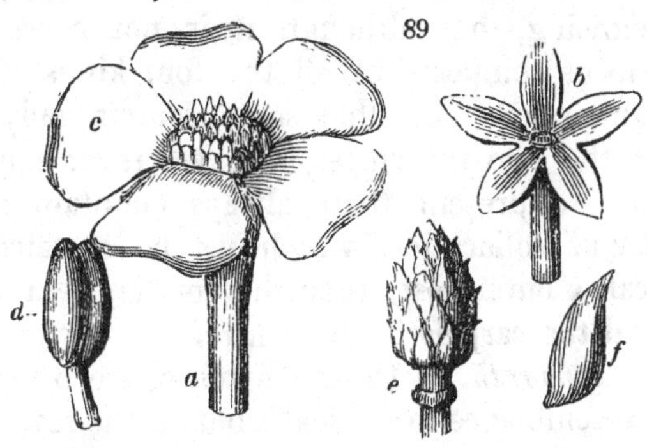

perianth consists of five parts, of a greenish yellow colour, and is sufficiently distinguished from the next whorl, to admit of its receiving a specific appellation; it is therefore termed the "calyx" (*b*); whilst its subordinate parts are called "sepals." The five parts which compose the next whorl are of a bright yellow colour, and are termed "petals" (*c*), or, collectively, the "corolla." The calyx rarely consists of more than one whorl of sepals, but the corolla is frequently composed of more than one. Next, within these, are several whorls of "stamens," one of which is represented at (*d*). These are the fertilising organs of the flower, composed of threadlike stems, surmounted by oval cells, or pouches, which contain a fine powder, named pollen. Lastly, we have several whorls of "carpels" (*e*), which are little ovate bodies, containing the "ovule," or young seed. The carpels, like the sepals, are not often ranged in more than one whorl, though they are so in this instance; but the stamens frequently occupy several. When the carpels adhere together, so as to form one mass, this is termed a compound "pistil;" but when they are distinct, as in the present case, each forms a separate pistil. Having given a general notion of the various parts of the flower, we must now enter a little more fully into a description of the several whorls, and mention some of the numerous modifications which they present; also premising, that although it is not necessary for flowers to be composed of all the four kinds of organs here enumerated, and that some contain only one or other of the two innermost, yet, wherever more than one kind are present, these always maintain the precise order of collocation, which we have stated above — the calyx outermost, then the corolla, next the stamens, and the carpels in the centre.

(93.) *Perianth.* — In the bracteæ, we often find a striking resemblance to the leaf; but in the several parts of the perianth, this becomes so much slighter, that in most cases the close affinity between these organs would scarcely be acknowledged, were it not clearly perceptible in some flowers; and also established by those cases of monstrous development, where the several parts of the perianth assume a leafy appearance. In many cases, and especially in monocotyledonous plants, the several whorls of the perianth so nearly resemble each other, that no distinction can be drawn between calyx and corolla, and the separate parts are described as "segments of the perianth." In those Dicotyledones where the perianth consists of a single whorl, it generally assumes the usual characters of a calyx; and is always so considered by most modern botanists, though Linnæus and others, have described it as a corolla, in many species where it happens to be coloured. Stomata exist both on the calyx and corolla, but more especially on the former.

(94.) *Calyx.* — Although the calyx very frequently "persists," — or remains whilst the fruit ripens, after the corolla has fallen, — it is in some instances very fugacious. The sepals frequently cohere by their edges into a tube, and the calyx is then "monosepalous," or "monophyllous," or more correctly "gamosepalous." In proportion as this cohesion extends from the base towards the apices of the sepals, the several modifications which it presents receive different appellations. It is termed "partite," when the cohesion extends but a short way; "divided," when it reaches about half-way up; "toothed," when it is nearly complete; and "entire," when the sepals are completely united to the very summit. In this last case, the number of the sepals can only be ascertained by their venation, each separate sepal being indicated by the position of its midrib; but in the other cases, which are most usual, the free apices of the sepals readily point out their number. Some sepals are so firmly united by their apex into one piece, that no separation takes place in this part, as the corolla enlarges. The calyx is then ruptured round the base, or transversely across the middle, and is thrown off in the form of a little cup, as in Eucalyptus (*fig.* 90.). When the cohesion is more perfect between some sepals than others, so as to form two lobes to the calyx, it is termed "lipped." An analogy is frequently maintained between sepals and the leaves, in such plants as bear stipules. This is indicated by the presence of little scales, resembling bracteæ, seated on the outside of a monosepalous calyx, and alternating with the sepals themselves, as in Potentilla (*fig.* 91.).

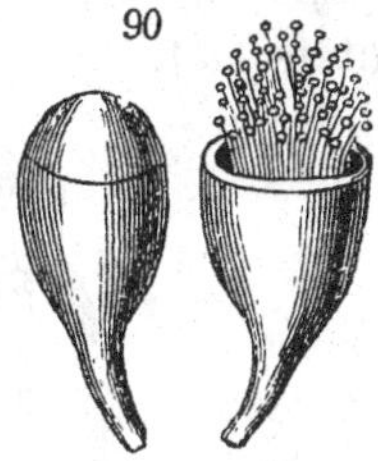

(95.) *Corolla.*— The petals are generally even less leaf-like than the sepals, more highly coloured, and more variously modified in shape. Like the sepals, they are either free, or cohere by their edges, forming a "monopetalous" corolla. In many cases, the petals may be divided into two parts — the "claw," which is analogous to the petiole of the leaf; and the "limb," which corresponds to the limb of that organ. By the cohesion of the claws, a tube is frequently formed, whilst the limbs continue more or less free, and appear as a border round the top of it. In some cases, the petals adhere at the base and apex, but are free in the middle, as in Phyteuma. An irregularity in the cohesion, produces a lipped corolla, as in the case of the calyx. We will here enumerate a few of the most important forms which the corolla assumes, the most remarkable of which are among such as are monopetalous.

1. *Regular monopetalous Corollæ.* — Where the several parts are symmetrically arranged round the axis, the forms are named after certain appearances which they are supposed to resemble; as the bell-shaped (*fig.* 92. *a*), funnel-shaped (*b*), salver-shaped (*c*), rotate (*d*).

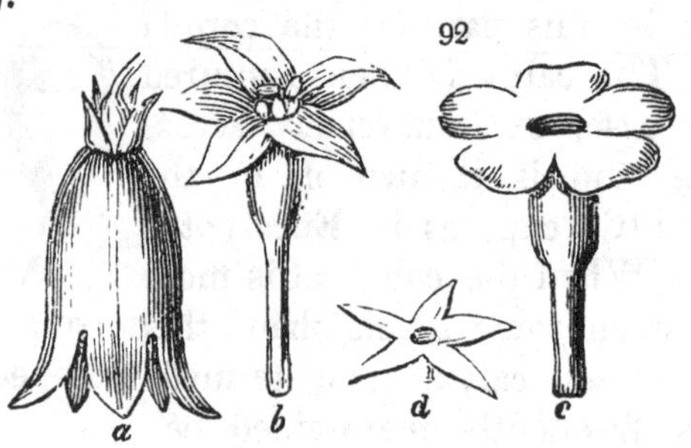

2. *Irregular monopetalous Corollæ.* — Where the petals cohere, but one part of the corolla is differently modified from another; as in the "lipped" or "labiate" flower (*fig.* 93.), which has two lobes forming the limb; and the "personate" flower (*fig.* 131. *a*), formed on somewhat the same plan, but where the mouth of the tube is closed. In these, and in other cases of irregular monopetalous corollæ, it is not always easy to distinguish the precise number of petals which cohere together, although we may generally do so by examining the venation, or by observing the apices of

93

the petals, which are free, and project beyond the margin.

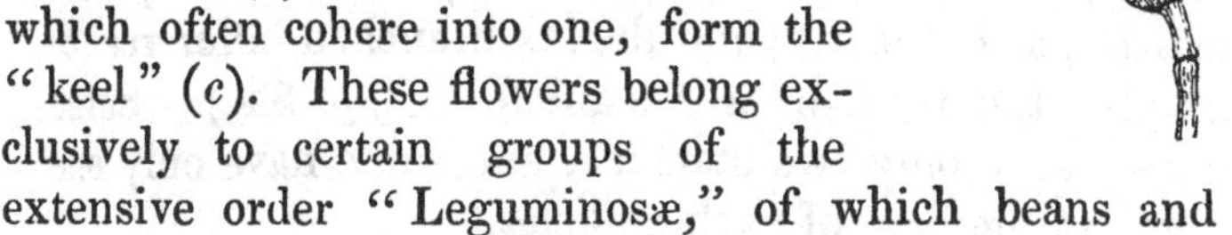

3. *Irregular polypetalous Corollæ.* — One of the most prominent of this class is the "papilionaceous" flower (*fig.* 94.), composed of five petals; which, however, are not always free at their base; but in a few cases cohere by their claws into a tube. The large single petal is termed the "standard" (*a*); the two lateral, the "wings" (*b*); and the two others, which often cohere into one, form the "keel" (*c*). These flowers belong exclusively to certain groups of the extensive order "Leguminosæ," of which beans and peas are familiar examples.

There is a vast variety among the irregular polypetalous corollæ, originating in peculiarity of shape, and in the proportion and numbering of the several parts.

(96.) *Glumaceous Flowers.*—The grasses (*Gramineæ*) and sedges (*Cyperaceæ*) have their flowers constructed in so peculiar a manner, that it will be necessary to describe them somewhat more particularly. Their perianth consists of membranous scales termed "glumes," which are referable to a modification of bracteæ, rather than of those more or less flaccid and foliaceous organs, which we have described as sepals and petals. In the example selected for *fig.* 95., there is a pistil (*a*), composed of an ovarium which contains a single ovule, and is surmounted by two stigmas. At the base are two scales. There are three stamens. These parts are included between two glumes (*b*), one of which is towards

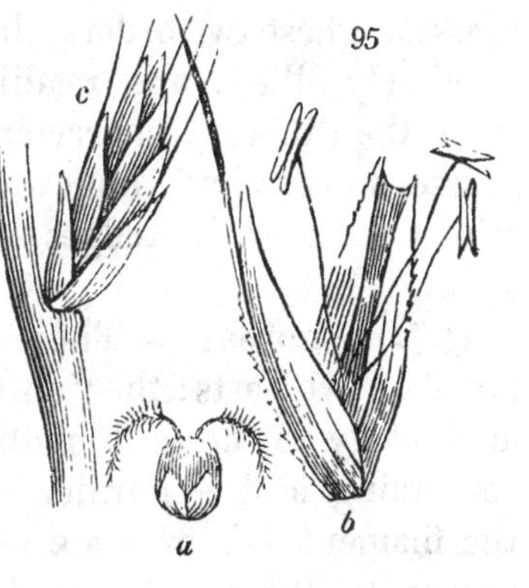

the stalk, or "rachis," on which the flower is seated; and this glume appears by its nervation to be composed of two united; this is further indicated by a little notch at its apex. The other, or outermost glume, is furnished with a bristle-shaped projection at the back, termed an "awn." Several of these flowers are closely ranged on opposite sides of a stalk, and form a "spikelet" (*c*), which is itself contained between two glumes at the base. When several of these spikelets are arranged alternately on the main rachis, they form a spike, as in wheat. In some examples, the flowers have three glumes. Some flowers are solitary, and on separate pedicels, as in the oat; and the lax branched inflorescence assumes the form of a "panicle" (*fig.* 84.). Some grasses have only two stamens, and some have only one glume at the base of each spikelet.

In the Cyperaceæ (as in *fig.* 96.) we have only one glume to each flower (*a*). The pistil (*b*) is inclosed in a membranous bag (at *a*), composed of two glumes united. The stamens are two or three, as also are the stigmas. The flowers of many of the Cyperaceæ are unisexual, and arranged in spikelets and spikes, much in the same way as in the grasses. These two orders, although so closely allied, are readily distinguishable; for besides the different character of their inflorescence, the grasses have round, hollow, and jointed stems (*culms*), whilst those of the sedges are more or less angular, and solid.

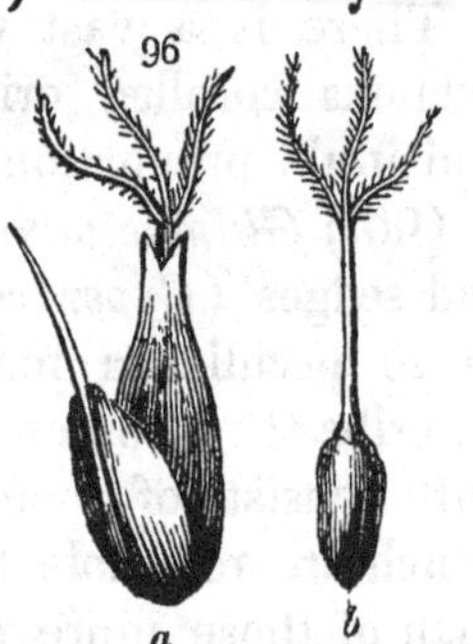

(97.) *Stamens.*—These organs are generally composed of two parts: the "anther" (*fig.* 97. *d*), which bears an analogy to the limb of the leaf, and is a sort of pouch containing a fine powder termed the "pollen;" and the filament (*e*) or stalk upon which it is seated, analogous to the petiole, or leaf-stalk. The latter part, however, is sometimes wanting, and then the anther is consequently sessile. Sometimes the filaments cohere, and form a tube round the carpels, and the stamens are then termed "monadelphous" (*fig.* 97. *a*). When they cohere into two separate bundles, they are said to be "diadelphous;" and when they appear in more than two, "polyadelphous." In some orders, but more particularly in the extensive order of the Compositæ, where this circumstance is universal, the filaments are free, whilst the anthers alone cohere, and form a ring round the pistil (*b*). This disposition of the stamens is termed "syngenesious." In some plants the filaments are dilated and closely resemble petals (*c*), to which organs they also frequently adhere through a greater or less extent.

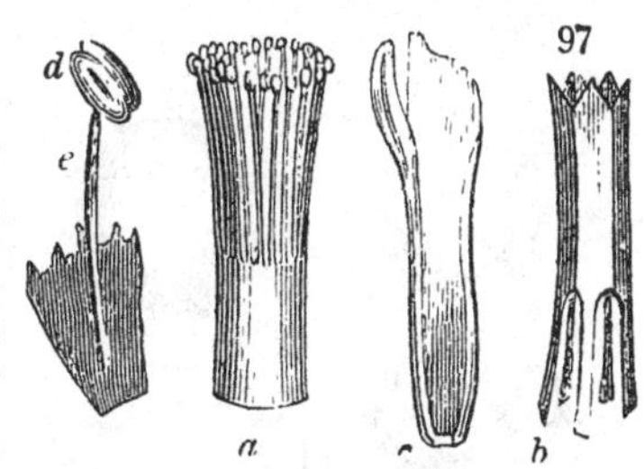

(98.) *The Anther* generally consists of two separate lobes or pouches, which contain the pollen (*fig.* 89. *d*); and this, when fully ripened, escapes through a fissure. When the fissure is closed, excepting at one extremity, the opening is a mere pore (*fig.* 98. *a*). In a very few instances the pollen escapes through vales, formed on the face of the anther (*b*). That part of the filament by which it is connected with the lobes of the anther, is termed the "connective;" and although more frequently obscure and of small dimensions, yet in some species it spreads, or branches laterally, and keeps the two cells wide apart (*c*). The cells themselves assume various appearances, and sometimes only one is perfected. In its earliest state, each is subdivided by a partition, which afterwards disappears; but in some cases it remains, and then each lobe contains two cells.

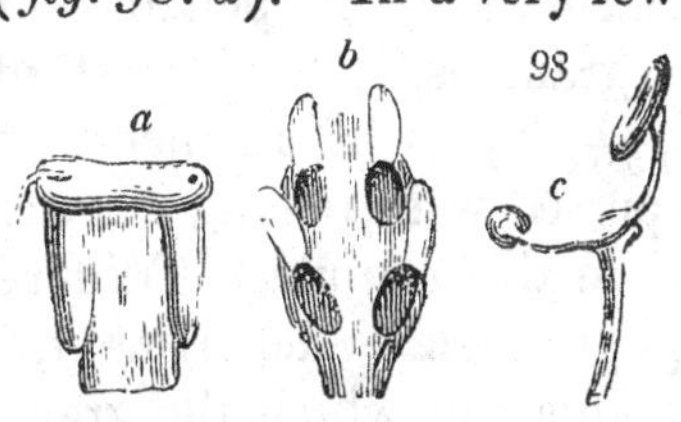

(99.) *Pollen.* — The grains of pollen (*fig.* 99.) are minute vesicles composed of one or two membranous coats, and are generally spherical or spheroidal, and often have determinate markings, warty projections, and

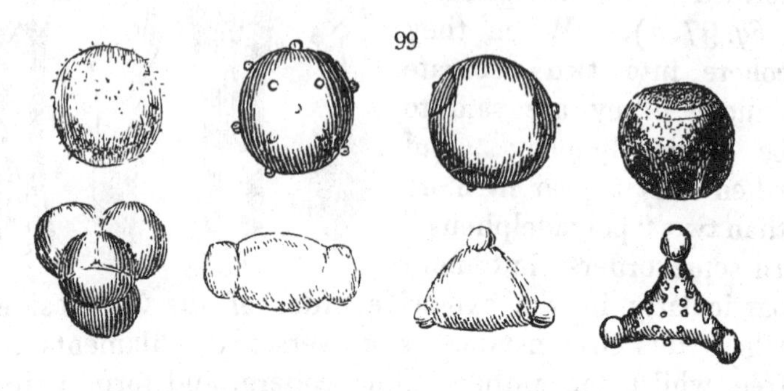

minute bristles upon their surface. Some of the largest grains do not exceed the $\frac{1}{360}$ or $\frac{1}{720}$ part of an inch in diameter; and in some species they are not so much as the $\frac{1}{2000}$. In several species, the grains approach a tetrahedral shape; others are very singularly modified, of which the few examples represented in the annexed cut may serve as a specimen. In some tribes of the remarkable order Orchideæ, the grains adhere together in waxy "masses," which fill the anthers. Each grain of pollen contains a quantity of minute "granules," the largest of which do not exceed the $\frac{1}{15000}$ part of an inch. These are occasionally interspersed with oblong particles, two or three times larger than the granules. We reserve further details for the physiological department, when we shall speak of the manner in which the grains act upon the stigma, in securing the fertility of the ovule.

(100.) *Pistil.* — The parts which compose the innermost whorl or whorls, are termed carpels, as we have already stated (art. 92.); and when they are not united together, each is also considered as a "pistil." This pistil, whether simple or compound, consists essentially of an "ovarium" or "germen," containing the young seed or "ovules;" and of a "stigma," or glandular summit, which is either seated immediately upon the ovarium, or on a sort of stalk, called the "style," interposed between them. The construction of the compound pistil will be more readily understood, by considering the manner in which the carpels themselves may be supposed to originate. Each carpel is an organ, analogous to a leaf folded inwards upon its mid-rib, so as to bring the edges into contact, which cohere and form the "placenta," and upon this the ovules are produced. In general, the carpels may be likened to a sessile leaf; but in a few cases they are furnished with a support (*thecaphore*) analogous to the petiole. When two or more carpels are placed closely in contact, and adhere together by their sides, the compound ovarium will contain two or more "cells" (*fig.* 100.) And if the styles and stigmas also cohere, the pistil will assume the appearance of a simple organ, although, in fact, compounded of two or more carpels. Where there is more than one row of carpels in the composition of a pistil, this will contain more than one tier of cells; as in the fruit of the pomegranate (*fig.* 101.).

100

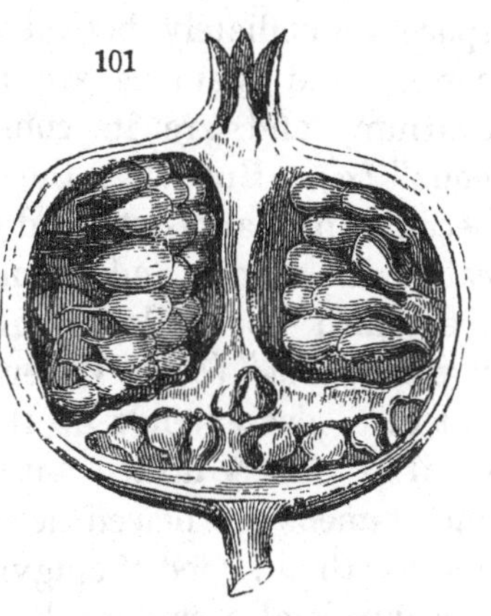

101

The *stigma* is variously modified in different species. It consists of vesicles of cellular tissue denuded of the epidermis, excepting in a few cases, where the thin pellicle which we have stated to form the outer skin of this investing organ, appears to cover it.

(101.) *Disk.*—The term "disk," is applied to a portion of the torus between the calyx and pistil, when it assumes a glandular, swollen, or fleshy appearance. This is always supposed to proceed from the abortion, or imperfect development of some of the petals and stamens. The disk, therefore, is not properly a distinct organ; but merely a modification of one or other of these. As connected with the development and modification of the torus itself, we may here describe three conditions of the flower, which are considered of the greatest importance in systematic botany, and which we will explain by referring to the annexed diagram (*fig.* 102.). When that part of the torus from

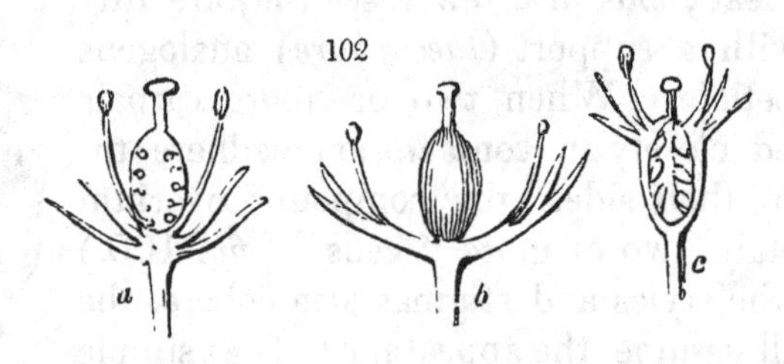

which the petals and stamens originate, is limited to the space immediately between the calyx and pistil: the corolla and stamens are necessarily seated below the ovarium, and are in consequence termed "hypogynous" (*a*). But when the torus is so developed, that it becomes partially extended over the inner surface of the calyx, the corolla and stamens appear to arise from, and are seated upon, this organ, and they are then termed "perigynous" (*b*). When the torus, modified as in the last case, also extends up the sides of the ovarium, the pistil is closely united with the calyx; and the corolla and stamens are placed near the summit of the ovarium, and are then styled "epigynous" (*c*). In this case, the ovarium is also said to be "inferior," with respect to the other parts of the flower, and these again are called "superior," with respect to it. In the perigynous and hypogynous corollæ, the reverse is the case, the ovarium being superior and the other parts inferior. There are a few other modifications which cannot exactly be referred to either of these three. In the white Water-lily (*Nymphæa alba*), the petals and stamens are attached to the sides of the ovarium, though the calyx is perfectly free. In the passion-flowers, the stamens adhere to the ovarium, and the petals to the calyx.

(102.) *Floral Modifications.* — As an illustration of these, we may state, that the orders of the class Dicotyledones, are thrown into four principal groups, two of which are characterised by the circumstances alluded to in the last article. The first of these, the Thalamifloræ, includes those flowers which have their several whorls detached, or not adhering together — each whorl occupying a distinct position on the torus, as in *fig.* 89. The separate parts of the several kinds of whorls, however, may or may not adhere together. This group can strictly include only hypogynous flowers. The next, or the Calycifloræ, includes those orders whose flowers have their petals and stamens adhering to the calyx, whether in the perigynous or epigynous form of the flower. In both groups, all the four floral whorls are almost universally present. Each, however, contains a few examples which cannot be separated from their congeners, but in which the petals are wanting, or are very rarely developed.

Of the two other groups, one is termed Corollifloræ, where the corolla is monopetalous, and the stamens adhere to the inside of its tube. This includes only hypogynous flowers. The last group is termed Monochlamydeæ, where the perianth consists of only one whorl, which is almost universally recognised as a calyx.

(103.) *Nectary.*—The word "nectary," is of very general application, and is used to express some peculiar modification in the sepals or petals, by which they assume an unusual form; but more especially, when there is some alteration of structure, by which they are wholly or partially converted into secreting organs, and exude a saccharine, glutinous juice.

(104.) *Æstivation.* — As the condition of the leaf whilst yet in the bud, is termed its vernation, so the manner in which the several parts of the flower lie folded in the flower-bud, is termed their "æstivation." Of this there are several kinds; the most important distinctions depending upon whether the edges of two contiguous sepals or petals meet without overlapping — when the æstivation is called "valvular" (*fig.* 103. V); or whether the one overlaps the other — when it is termed "imbricate" (*fig.* 103. I). The various modifications to which the æstivation is subject, is readily seen, by making a transverse section through the flower-bud. Thus, the "conduplicate" (*fig.* 104. C), is where the edges in the valvular æstivation, are rolled inwards beyond the line of contact. The "contorted" or "twisted" æstivation (T), when the parts of an imbricate æstivation are so curved, that each is partially wrapped round one, and at the same time is partially enveloped within another. These examples are sufficient to afford a general notion of this phenomenon.

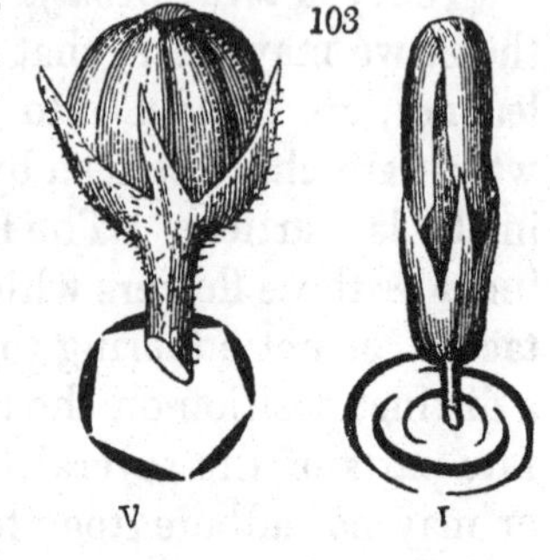

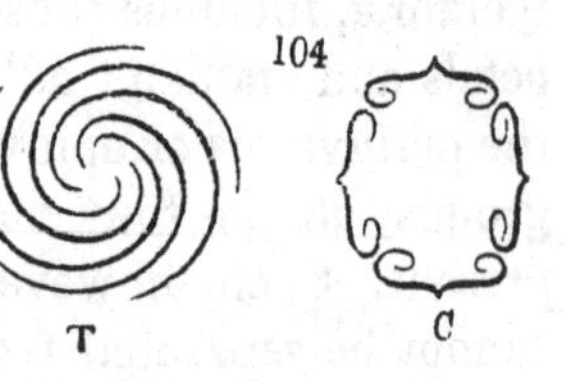

CHAP. V.

REPRODUCTIVE ORGANS — *continued.*

FRUIT — PERICARP (105.). — FORMS OF FRUIT (108.). — SEEDS (109.). — EMBRYO (111.). — REPRODUCTION OF CRYPTOGAMOUS PLANTS (114.).

105.) *Fruit.* — IMMEDIATELY after the flower has become fully expanded, several of its parts begin to decay; but the ovarium, sometimes the calyx, and other parts continue to grow, and ultimately assume a very different appearance from what they possessed in the flower. This altered condition of these parts is termed the "fruit." In many cases, the fruit is not ripened unless the ovula are subjected to the fertilising influence of the pollen; but if this process be completed, then these bodies undergo certain remarkable changes, and pass to the condition of "seeds." Certain fruits, however, will ripen freely enough, although they produce no seed, as some varieties of oranges, grapes, pineapples, &c.

(106.) *Pericarp.*—The part of the fruit immediately investing the seed, and which originally formed an ovarium, becomes the "pericarp." When the carpels are separate, the fruit is termed "apocarpous;" but when composed of several adhering carpels, it is said to be "syncarpous." The pod of a common pea, is a familiar example of a simple pericarp, with a structure not very dissimilar to that of a leaf folded longitudinally inwards, with the seeds attached along the margins, united and forming a swollen placenta. De Candolle has given a figure, in his "Memoir on

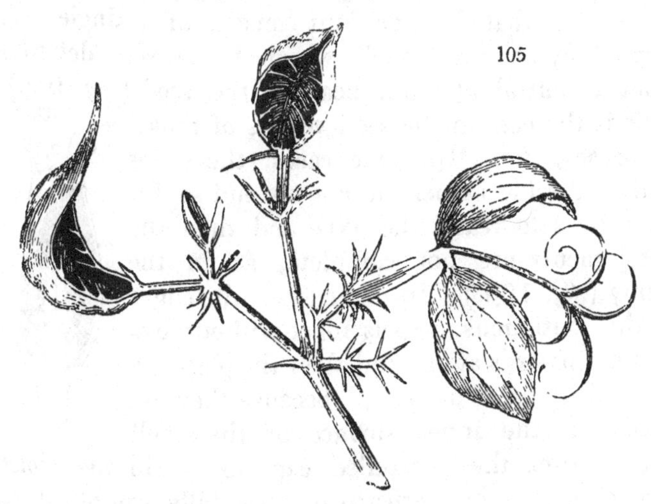

the Leguminosæ," of a monstrosity, where the pericarps

have manifested a decided tendency to develop in the form of leaves, and where the position of the ovules is marked on their edges by small projections (*fig.* 105.).

106

If we suppose five carpels, formed on the same general principle as that of the pea-pod, to be arranged round an axis, and to be enveloped in a mass of pulpy matter, contained in a swollen calyx (as in the apple blossom), we have such syncarpous fruits as apples, pears, &c. (*fig.* 106.).

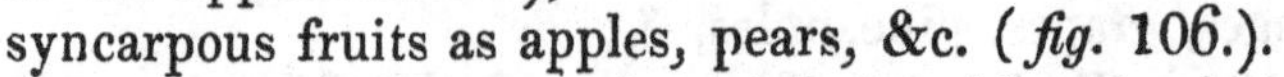

A multitude of examples might be adduced, where the compound structure of the pericarp is easily referable to an aggregation of several carpels. In such cases, each carpel forms a distinct "cell;" and the wall of separation between two contiguous cells, is termed a "dissepiment" (*fig.* 107.). There are, however, many pericarps, which, in their nascent state, possess this structure, but become further modified as they ripen, by the rupture and subsequent obliteration of the dissepiments; at the same time the placentæ coalesce round the axis, so that the ripe fruit consists of a single cell, formed by an outer shell, which is entirely detached from a central placenta bearing the seed (*fig.* 108.). This is the case in the seed-vessels of pinks, primroses, &c. In some cases, the edges of the adhering carpels do not extend so far inwards as to reach the axis, and then the dissepiments are not complete, as in the poppy (*fig.* 109.). In other cases, the edges of the contiguous carpels meet without extending inwards at all, and then the placentæ are said to be "parietal," because they are placed on the inner surface of the shell which forms the one-celled capsule, as in the violet (*fig.* 110.). The pericarp is essentially composed of three parts, analogous to those in the leaf—two skins, and

107

108

the cellular matter between them. The outer skin forms the "epicarp," the inner the "endocarp," and the intermediate portion is the "sarcocarp." In many pericarps, these parts are not well defined; but in such as are fleshy, as in the stone-fruits, peaches, plums, &c., it is the endocarp which develops into the "stone," the epicarp forms the "skin," whilst the sarcocarp becomes the delicious and edible portion of the fruit.

109

(107.) *Dehiscence.* — When the ripened pericarp divides spontaneously, in any definite manner, it is said to be "dehiscent," and the line of division is termed the "suture," whilst the separate parts are called "valves" (*fig.* 111.). In general, the suture tallies either with the adhering edges of the carpels, or with a line parallel and midway between them, in the position of the midrib or nerve of each carpel. In the former case, the dehiscence is termed "septicidal" (*a*), as in the *Colchicum autumnale;* and in the latter, which is the most usual, "loculicidal" (*b*), as in the tulip. In a few plants, as in the common pimpernel (*Anagallis arvensis*), the suture is transverse to the lines formed by the edges of the carpels; such a pericarp is termed a "pyxidium" (*fig.* 112.). In some cases, the dehiscence is so limited, that it merely forms pores or small valves, at the extremities of the pericarp. In many pericarps there is no particular line of suture; but they rupture irregularly, to permit the escape of the seed; or else they decay and gradually rot without bursting.

110

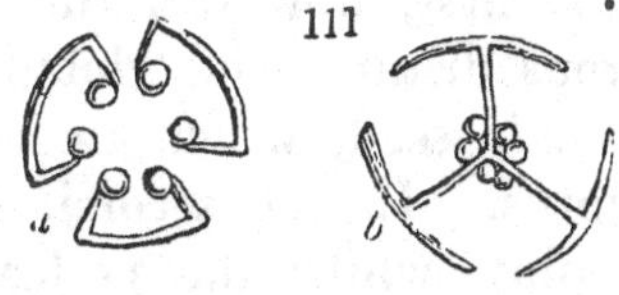

112

(108.) *Form of Fruits.* — It would be impossible in this treatise to enumerate the vast variety of forms and characters which different fruits present. Some are soft and pulpy; others are very hard, woody, dry, or membranaceous. It is sometimes one part, and sometimes another, of the inflorescence, which becomes developed into a succulent and nutritious form, in different fruits; and a casual observer might easily overlook these distinctions, in the general resemblance which they bear to one another (*fig.* 113.).

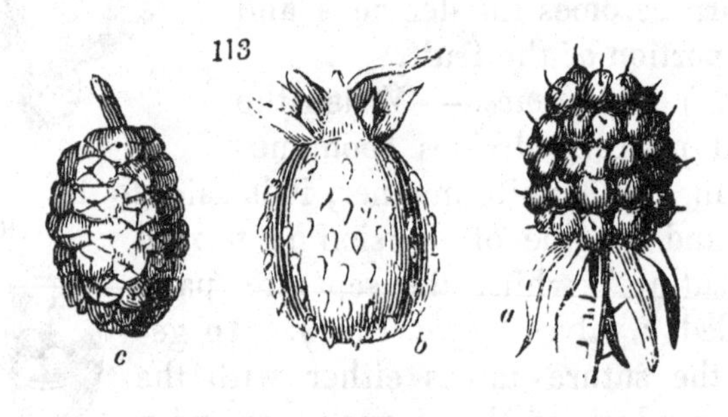

The raspberry (*a*), the strawberry (*b*), and perhaps the mulberry (*c*), may be mentioned, as bearing a considerable general resemblance to each other. In the first, however, the juicy part consists of numerous distinct and globular pericarps, each enclosing a single seed, which are seated on a spongy unpalatable torus. In the second, it is the torus which becomes pulpy, whilst the pericarps remain dry, and are scattered over its surface in the form of little grains, commonly considered as naked seeds. In these two cases, the fruit is the produce of a single flower; but in the mulberry, the structure is altogether different. This tree is monœcious; and the small fertile flowers — or such as contain pistils, and no stamens — are disposed in a dense spike. It is the calyx of each flower which becomes succulent, and thus the fruit is made up of the aggregate mass of these altered calyces, each of which invests a dry pericarp, containing the seed.

We shall very briefly notice a few of the most important forms which fruits assume, but cannot pretend to enter into any details on so extensive a subject. Dr. Lindley's "Introduction to Botany" may be advantageously consulted for further information, and Gærtner's invaluable works for the fullest details.

Simple Pericarps.

1. *Follicle.* — Where the pericarp is dry, and dehiscent only along the suture formed by the union of the edges of a foliaceous carpel, it may be considered as composed of a single valve: as in the monkshood (*Aconitum napellus*), and larkspur (*Delphinium consolida*, *fig.* 114.).

2. *Legume.*—This form is familiarly illustrated in the pericarps of peas and beans. In many cases, it presents a near approach to the leafy structure, and may be considered as a modified condition of the leaf, folded longitudinally on its midrib, with the edges adhering, and forming a suture (*fig.* 115. *a*). Another

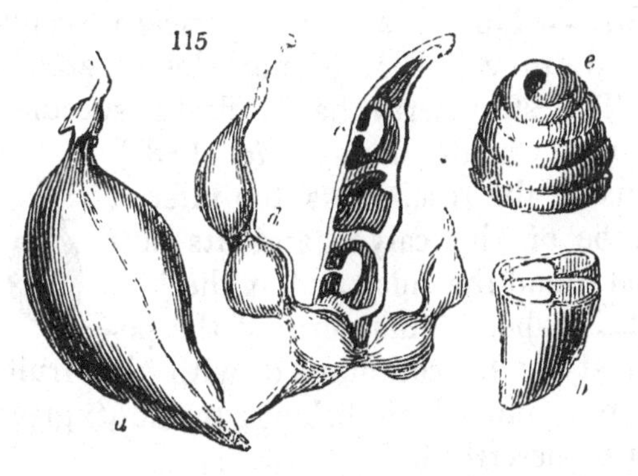

suture is also formed along the midrib or dorsal nerve, so that the legume separates into two valves. In some species, however, the sutures are so firmly closed, that the legume becomes indehiscent. Its varieties are very numerous. In the genus Astragalus, it is divided into two spurious cells (*b*), by the back of the legume becoming doubled inwards until it reaches the placenta. In some cases, the legume is divided by transverse partitions (*c*), formed by the agglutination of the opposite parietes, so that each seed appears to be contained in a separate cell; and in some cases the pericarp is pinched between each seed, so that the sides nearly meet, when it is termed "lomentaceous" (*d*). In some cases it falls to pieces at these transverse contractions, and breaks up into as many detached cells as there are seeds. In the genus Medicago, the legume is curiously twisted in a spiral manner, and somewhat resembles a snail-shell (*e*).

3. *Drupe.* — This form may be illustrated by the plum, cherry, and other stone-fruits, where the pericarp has a thickened and pulpy mesocarp, with a stony endocarp. It contains two seeds in the early state; but one of them is most frequently abortive, and withers completely before the fruit is ripe. The numerous small drupes, or "drupels," of the raspberry, and other Rubi, are closely aggregated on a spongy convex torus (*fig.* 113. *a*).

4. *Nut.* — This is a bony pericarp, containing a single seed, to which it is not closely attached (*fig.* 116.). The strawberry has a fleshy succulent torus, covered with small nuts (*fig.* 113.). The torus of the rose, coats the interior of the tube of the calyx, and its nuts are placed round the sides and at the bottom of this tube. This form of the pericarp must not be confounded with the fruit usually called a nut, and which belongs to the "glans," presently to be described.

116

Pericarps simple by Abortion.

5. *Cariopsis.* — This pericarp is a thin, dry, and indehiscent membrane, closely investing, and in-

deed adhering to, the seed—as in corn, and other Gramineæ. As these pericarps bear two or three stigmas, the seed is probably simple by abortion, and therefore the fruit, strictly speaking, is compound.

117

6. *Akenium.*—This may be considered as a cariopsis, with the superaddition of the calyx, adhering to the pericarp, and forming a single skin round the seed — which, in this case also, is simple by abortion. The fruit of the "Compositæ" are formed on this plan (*fig.* 117.).

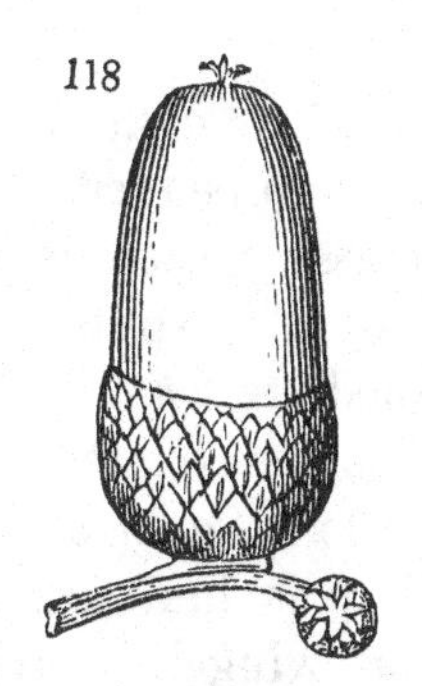

118

7. *Glans.* — Acorns (*fig.* 118.), hazel nuts, and chestnuts, are examples of this form. The base of the fruit is enveloped by an involucrum, which at first contains several flowers, but one of them alone perfects its seed. The pericarp is tough or woody, indehiscent, adhering to the perianth, one-celled by abortion, and containing one or more seeds.

8. *Capsule.* — This is a very general term, for dry fruits composed of two or more carpels, variously combined and modified.

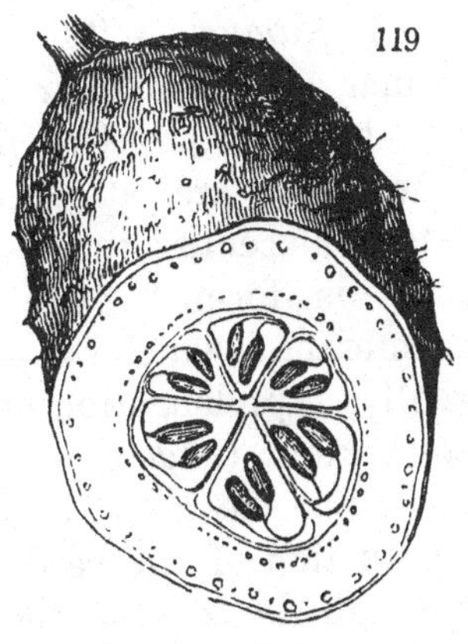

119

9. *Gourd.* — The carpels are not complete, but united by their edges so as to form a single cell with parietal placentæ. The pericarp is thick and fleshy, with the outer coat hard (*fig.* 119.).

10. *Berry.* — This term is applied to very liquid fruits, which are covered with an indehiscent skin, as the grape, gooseberry, and others. In the gooseberry the carpels are incomplete, and form one cell with parietal placentæ (*fig.* 120. *a*); and the calyx

adheres to the pulpy pericarp; but in the grape (*fig.* 120. *b*), the calyx is free, and forms no part of the fruit; the carpels are complete, and the placentæ central.

120

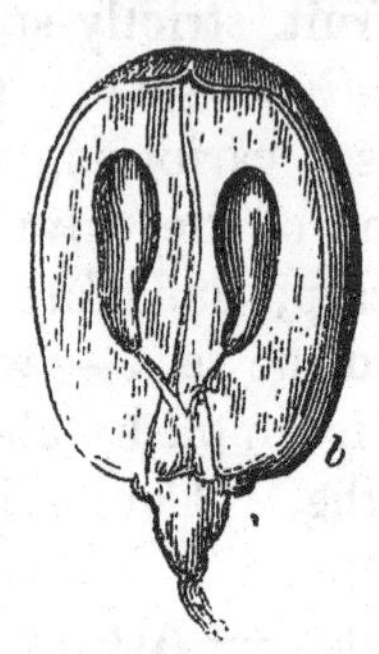

11. *Pomum.*—Several membranous, or bony carpels, are embedded in a fleshy mass, which is the swollen calyx. Apples (*fig.* 106.), medlars (*fig.* 121.), quinces, &c., are examples.

121

12. *Samara.* — The pericarp is here extended into a flat wing-like appendage, as in the sycamore (*fig.* 122.) and ash; the fruit of which trees is commonly termed a "key."

13. *Siliqua.* — This is the name given to the bilocular and bivalvular seed-vessels of the Cruciferæ. The seeds are attached to lateral placentæ; the dissepiment is formed by a thin membrane, which is apparently a prolongation of the inner skin (*endocarp*) of the two carpels (*fig.* 123.).

122

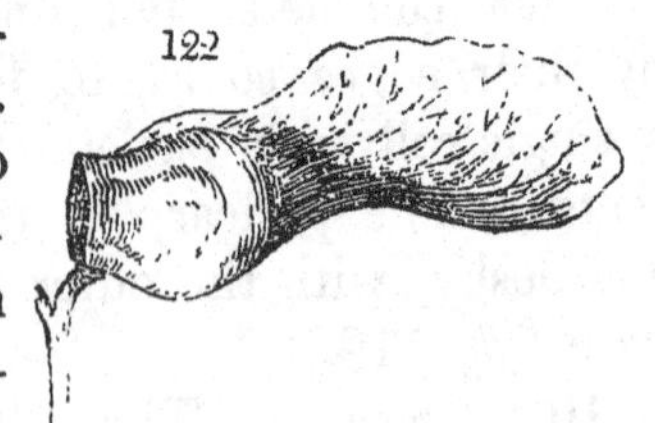

(109.) *Seed.* — It would be impossible to obtain a just notion of the seed, without first tracing the ovule through the several alterations which it undergoes, after it has been subjected to the fertilising influence of the pollen; but, as such details are more especially connected with the physiology of our subject, we shall for the present confine ourselves to a few general observations on the ripe seed. Every seed is attached to the placenta, by what is termed a "funicular, or umbilical cord;" and when the seed has fallen from the pericarp, it is marked by a scar or "hilum," at the place where this cord was attached to it. In very many cases, this cord is small, and scarcely distinguishable, but in some it is well marked; and in the genus Magnolia, when the pericarp bursts, the seeds hang out for some time, and to a considerable distance, by means of their umbilical cords, before they become detached and fall to the ground (*fig.* 124.). In a few plants, the funicular cord is unusually developed; and, rising round the seed, forms a distinct skin or covering to it, termed an "arillus." The nutmeg (*fig.* 125.) is thus enveloped by an arillus, which is the "mace" of commerce. In the spindle-tree (*Euonymus europæus*), the seeds are invested by an arillus, of a fleshy consistency, and bright scarlet colour.

124

In its ripe state every seed is essentially composed of an outer skin, or "spermoderm," and a "kernel" within it. The spermoderm, however, is not a distinct organ, but is rather the dry and exhausted remains of two or more coats, with which the embryo was invested in its earliest state, but which have ultimately united, and form a single skin on the ripe seed. The kernel consists of the "embryo;" and, in many cases, also contains a peculiar substance termed the "albumen," which is a nutritious matter secreted for the use of the embryo, and is either of an oily, farinaceous, or hard and horny, consistency. This substance is always wholesome; and in many seeds, especially in corn, forms an important article of human food. In some cases, the embryo is completely invested by the albumen, as in the cocoa-nut; in others it is only partially embedded, as in wheat and other corn (see *figs.* 23. and 25.). In a multitude of seeds, however, there is no trace of this substance, in a detached form; but then we often find the cotyledons themselves much swollen, and abundantly supplied with a similar material. This is the case in peas and beans, whose cotyledons are very large, and contain a nutritious material, which serves to develop the young plant in the early stages of its growth. Some few seeds, as the orange, contain more than one embryo; a fact which has been considered analogous to the phenomenon of double fruits, and to be explained on the supposition that two or more ovules have adhered together in the earliest state of their development.

125

(110.) *Forms of Seeds.* — The forms which seeds assume are very various, and their surface is either smooth, rough, or, in some cases, furnished with peculiar downy or membranous appendages. The various appendages, however, which assist the dissemination of the seed, are more frequently attached to the pericarp; and afford abundant instances of an adaptation of means, admirably calculated to secure the end for which the seed is destined — the preservation of the species upon the earth.

(111.) *Embryo.* — We have already described (arts. 34, 35.) the two principal distinctions, which subsist between the embryos of flowering plants, and which essentially separate them into two great classes. To those remarks, we may add the following:—The embryo may be either straight or curved; placed in the centre of the albumen, where this substance exists in a separate form, or else laterally disposed with respect to it. The parts of which it is composed are, 1. The "radicle," which is the conical extremity, afterwards developed into a root; and, 2. The "plumule,"—consisting of the "cotyledon or cotyledons," and the "gemmule," or first leaf-bud, which is afterwards evolved in the form of stem and leaves.

The position of the embryo is determined by the direction of its radicle, the point of which is constantly turned towards the "foramen,"—a small hole pierced through the outer coat of the seed, and of which we shall speak more particularly hereafter. Now, the position of the foramen varies with respect to the hilum, and may be either on the opposite side, or placed near it, on the same side of the seed. The radicle will, consequently, either point from or towards the hilum, and the embryo become "inverse" (*fig.* 126. *a*) or

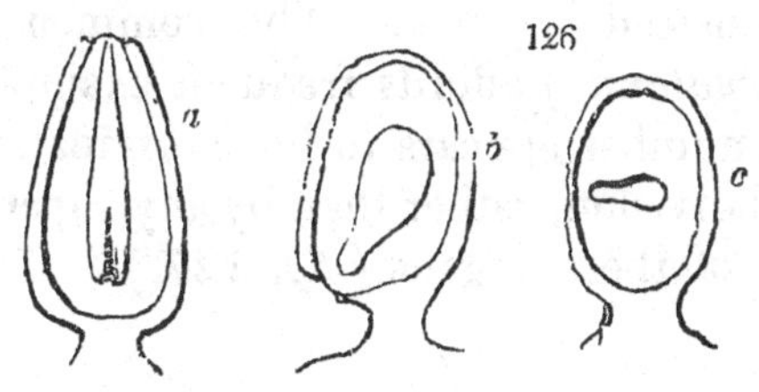

"erect" (*b*) accordingly; or the embryo may lie "transverse" (*c*), when the apex is on one side of the seed, and the radicle cannot be said to point either towards or from the hilum. Some authors, however, make the direction of the embryo to depend also on the position of the seed itself, which may be either erect or pendent within the pericarp; but this is a circumstance which can merely modify the direction of the embryo with respect to the pericarp, and not with respect to its position in the seed.

(112.) *Cotyledons.*—In many plants, the cotyledons have comparatively little resemblance to leaves, but in others they alter their character very considerably after germination has commenced; they then become green, and expand in a form which closely resembles the ordinary leafy structure. Some cotyledons, however, whilst still in the seed, have the appearance of miniature leaves, are extremely thin, and delicately veined (*fig.* 23. *a*); and no one could for a moment consider them in any other light, than as these organs in a young and undeveloped state. In many Dicotyledons, the embryo is a cylindrical body, with nothing more than a notch at one end, indicating the position of the cotyledons; but, in a few species, there is no appearance of any division, and then it is presumed that the cotyledons adhere together; or rather, if we judge from analogy, that they are entirely abortive. Their stem consists merely of a slender filament which twines itself round other plants, from which it extracts its nutriment by means of suckers provided for this purpose.

Here and there, we often find a young plant of several dicotyledonous species, which have three, or even more cotyledons, instead of two. The common sycamore (*Acer pseudoplatanus*) affords frequent examples, where this unusual number appears to have originated in some process of subdivision, rather than by any supernumerary development of these organs (*fig.* 127.). These devi-

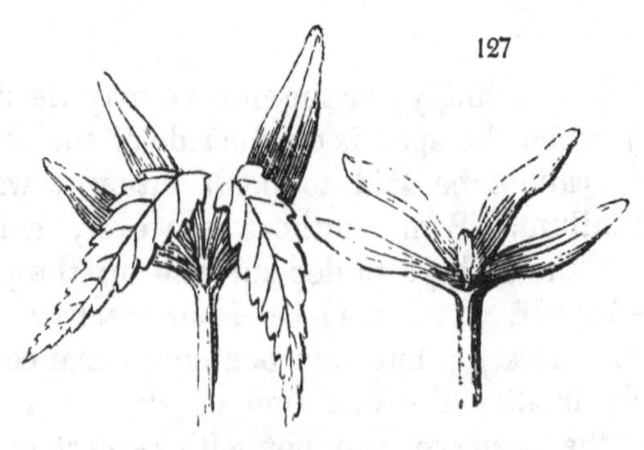
127

ations from the usual character, in species where the cotyledons are most frequently two in number, may serve as a connecting link between them and plants of the coniferous tribes (the fir trees), which possess several cotyledons.

An attempt has been made, to establish an affinity between the embryonic structure of dicotyledons and monocotyledons, by supposing the single cotyledon in the latter class, which completely envelopes the rest of the embryo, to be in reality compounded of two cotyledons, united by their edges into one mass. In some cases this occurs in dicotyledons; and the annexed figure (128.) represents a monstrosity, observed in a young plant of the sycamore, which exhibits an approximation to the condition of a monocotyledon, at the commencement of its germination: the two cotyledons having adhered by one of their edges nearly throughout their whole lengths.

128

In all monocotyledons, it is more difficult to determine the several parts of which the embryo is composed, than in dicotyledons. It generally consists of a nearly cylindrical fleshy mass, without any external traces of organisation; but if it be cut longitudinally, the position of the radicle and the gemmule may then be seen, traced by a faint outline, indicative of a separation in the substance of the embryo (*fig.* 25.).

(113.) *Reproductive Organs of Cryptogamic Plants.*—The sporules mentioned in art. 36. are contained in peculiar cells placed on the surface, or embedded in the substance of the plant, among the cryptogamic tribes. Among the higher families of this class, the cells assume a distinct capsular form, termed "theca" (*fig.* 129.), which has various characters, in the ferns (*a*), Equiseta (*b*), mosses (*c*), &c. The cells, or cases which contain the sporules, among the inferior families of this class, are more simple in their structure, and often re-

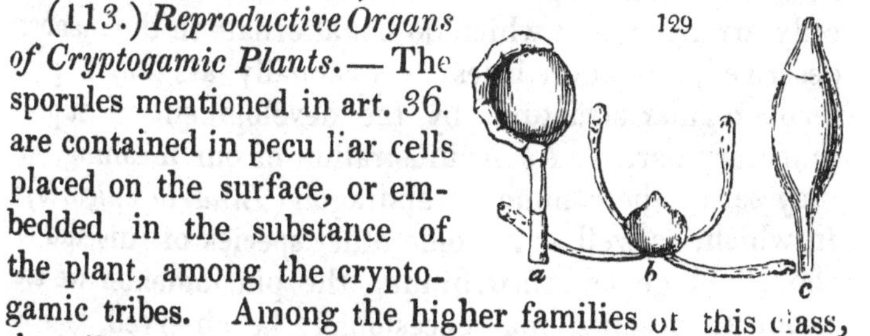

129

semble short closed filamentous tubes, or sacks (*fig.* 130.), which ultimately discharge their contents by the rupture of one of their extremities.

CHAP. VI.

MORPHOLOGY.

ABORTION (115.). — DEGENERATION (116.). — ADHESION (118.). — SUPERNUMERARY WHORLS (119.). — NORMAL CHARACTERS (120.). — SPIRAL ARRANGEMENT OF FOLIACEOUS APPENDAGES (121.). — TABULAR VIEW OF VEGETABLE ORGANISATION (123.).

(114.) *Morphology.* — It is an observed fact, that the subordinate parts which make up the floral whorls of very many plants, are symmetrically arranged round the axis, and that the parts of each separate whorl are placed alternately with those of the contiguous whorls. Connected with these facts, it has been remarked, that the flowers of certain species, whose parts are not symmetrically arranged, and which do not alternate in the manner described, do nevertheless occasionally assume a perfectly regular structure, by the development of supernumerary parts. As an illustration of our meaning, we may select the common snapdragon (*Linaria vulgaris*); in which, as well as in some other species of this and of the allied genus Antirrhinum, the phenomenon we are about to describe may occasionally be observed. The common form of the flowers of this plant is termed "personate" (*fig.* 131. *a*); the corolla is monopetalous, and divided into two large lobes, closed in front, and presenting somewhat the appearance of an animal's face. The upper portion of the corolla is prolonged backwards, into a tubular "spur;" it contains four stamens, arranged in pairs of unequal length (*didy-*

namous): the calyx is subdivided into five segments, indicating the adhesion of as many sepals; the pisti

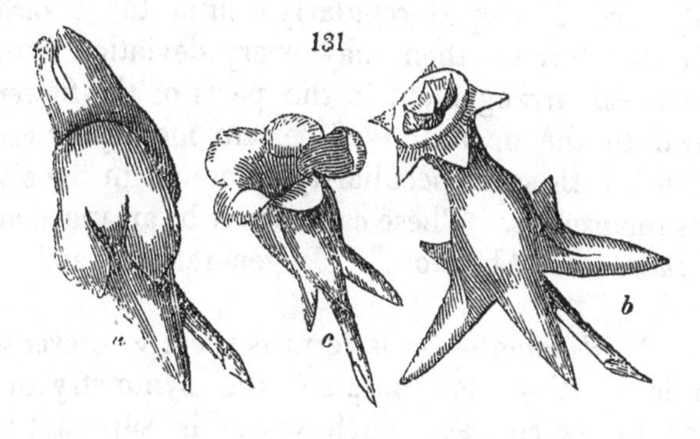

is a two-celled capsule, with the seeds arranged on a central placenta. In short, the flower is highly unsymmetrical and irregular, in all its parts. Now, there is an interesting variety of this plant, termed "*Peloria*," in which the corolla is strictly symmetrical, consisting of a conical tube, narrowed in front, and elongated behind into five spurs (*b*). It contains five stamens of equal length. In this state, therefore, we have a flower composed of five sepals, adhering through a considerable portion of their length, constituting a five-toothed monosepalous calyx; five petals, adhering into a monopetalous corolla; five stamens; but a pistil which is composed of only two carpels, as in the irregular flowers. The three first whorls are therefore strictly symmetrical, and the parts are also arranged in an alternating order round the axis. It should seem, that the ordinary irregularity of this flower is somehow connected with the disappearance of the fifth stamen, involving a partial suppression, as well as modification, of four of the petals. Other specimens may be seen in every intermediate condition, between the regular and irregular forms here described; some having two, others three or four spurs, to the corolla (*c*). If we connect these and similar facts, with the observations already detailed, viz. that the subordinate parts of the flower-bud are analogous to those which compose the leaf-bud, and consequently that all these parts are only analogous to so many leaves, which under other circumstances would have developed regularly round the branch on which they grew—then may every deviation from the symmetrical arrangement in the parts of the flower, be ascribed to the operation of certain modifying causes, connected with some peculiarity, inherent in the several species themselves. These causes may be arranged under the heads of "Abortion," "Degeneration," and "Adhesion."

(115.) *Abortion.*—This term is used, wherever some organ is wanting, to complete the symmetry of the flower; in which case, such organ is supposed to lie dormant under ordinary circumstances, though capable of development under other and peculiar conditions. As the latter are of accidental occurrence, they only give rise to those various monstrosities, or deviations from the ordinary form, which are frequently (as in the case of the Linaria above mentioned (art. 114.) so valuable), in determining what is considered to be the "normal" structure, or regular condition, to which various unsymmetrical flowers may be referred. Portions of the inner whorls are more often abortive than those of the outer; and thus the number of carpels is far less frequently in accordance with the normal structure, than the number of the stamens. All unisexual flowers, may be considered as resulting from the complete abortion of one or other of the two innermost whorls.

(116.) *Degeneration,* is where the abortion of an organ is not fully completed, but where it has become imperfectly developed, or very differently modified from its usual state. In many instances, we find certain anomalous appendages, which occupy the place of some of the subordinate parts belonging to one or other of the floral whorls, and which are consequently considered as a monstrous or incomplete state of those parts. Perhaps the stamens are more especially subject to this condition of degeneracy than any other organs. They frequently assume the form and structure of secretory glands, and of various processes and appendages, of an anomalous character. In many cases, the parts which have degenerated from their usual condition, assume a highly developed structure, and become more leaf-like. Thus, we find double flowers are often formed by the stamens having put on the appearance, and all the characters of petals,—organs which are usually of larger dimensions, though of inferior importance in the floral economy. In some plants, as the common white Water-lily (*Nymphæa alba*), the transition from the character of a petal to that of a stamen, is so very gradual (*fig.* 132.), through successive whorls of

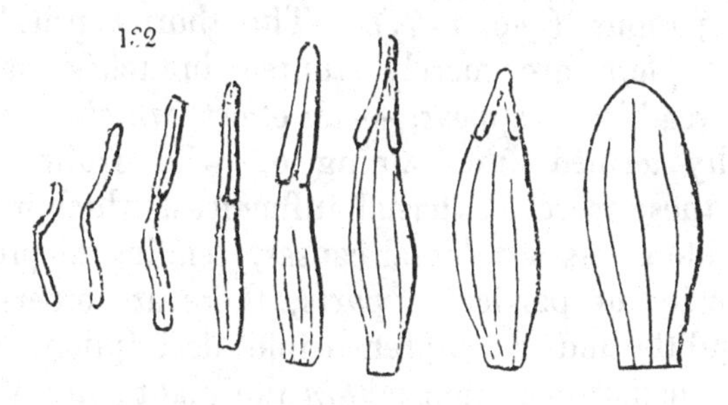

these organs, that it is hardly possible to determine where one set begins and the other terminates.

(117.) *Causes of Abortion and Degeneration.*—An inquiry into the causes of abortion and degeneration, more properly belongs to our physiological department, but may as well be alluded to in this place. The partial or total abortion of certain organs, is very frequently occasioned by accidental circumstances — from some impediment thrown in their way, from a deficiency of light in a particular direction, and many other *external* causes. In these cases, when the influence is removed, the suppressed organ will sometimes appear, and assume its proper character. Thus, in trees, it seldom happens that all the buds generated in the axills of the leaves, are developed into branches; but many of them remain dormant, especially about the lower parts of the stem; and it is not until a better supply of light and air is afforded them by the pruning knife, that they are enabled to grow. Sometimes the development of an organ is impeded or prevented, by the want of a sufficient supply of nutriment; and this often arises from the abstraction of what was naturally destined for it, by the more vigorous growth of some neighbouring portion. Hence the different characters which distinct individuals of the same species assume, depend upon the various degrees of influence which those and many other external circumstances have upon them. From such causes as these, we find the leaves of a tree gradually dwindling into membranous scales; the calyx of the florets in the Compositæ becoming a downy pappus (*fig.* 117.). The thorny prickles of the wild plum are merely stunted branches, and by culture readily disappear, — an effect which Linnæus fancifully termed, the taming of wild fruits. But besides these merely external influences, which may all be considered as accidental causes, tending to produce the abortion of particular parts, there are others of a more subtle and incomprehensible description, which are in constant operation *within* the plant; and which, acting from the very earliest periods in which certain organs begin to develop, tend to suppress or modify them; and thus produce that infinite diversity of forms and characters, which we find even among those which are destined to perform the very same function. And sometimes the altered organs are so far changed from their original character, as to become adapted only to serve some new secondary purpose, distinct from that for which they were primarily intended. Thus, the spines of the common furze (*Ulex europæus*), are merely modified leaves. In the common berberry (*Berberis vulgaris*), the transition may be readily traced (see *fig.* 68.).

(118.) *Adhesion.*—If to the operation of the two causes already noticed, we add the "adhesions," which take place between the contiguous parts of similar or different organs, we introduce a third cause, in very general operation, which serves to modify the normal condition of the several parts of the separate whorls. For example, the *Rhlox amœna* has a monopetalous tubular corolla (*fig*. 133. *a*), expanding into a flattened

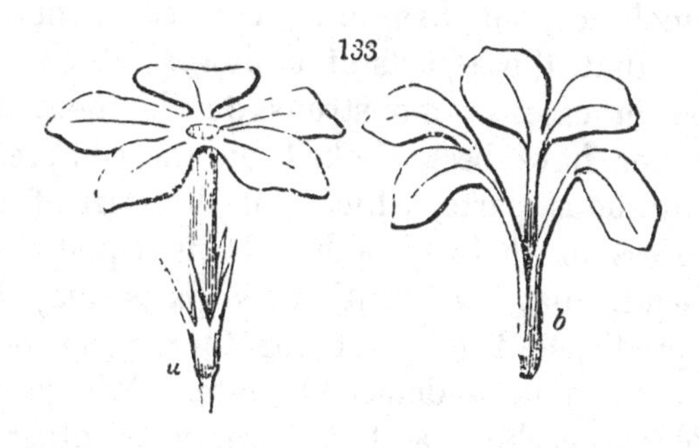

border at the summit, and forming what is called a "salver-shaped" flower. But a monstrosity of this plant has been observed, where the corolla is split up into five distinct petals, resembling those of a pink (*Dianthus*). This shows us, that the ordinary monopetalous condition of the corolla in this flower, has resulted from an adhesion of the five subordinate parts of which it is composed; and some blossoms have been found, in which this adhesion has only taken place partially, some of the petals being cemented half-way up the tube, whilst others adhere nearly throughout its whole length (*b*).

Not only may the several parts of the separate whorls contract adhesions of these kinds, but two or more of the whorls may be grafted together, throughout a greater or less extent.

The causes here enumerated, as modifying or disguising the several parts of which flowers are composed, are brought into operation at such early stages of their development, that it is very seldom we can trace the successive steps by which the metamorphosis has been effected. In many cases, however, we find the number of ovules in the ovarium, far exceeding the number of ripened seeds in the pericarp; and the obliteration of those which have become abortive, may be sometimes traced to the circumstance of there having been more ovules originally formed than could possibly be contained, as ripened seeds, in the pericarp, which would be too small to hold them all. It is easy, therefore, to conceive, that those parts of a flower which are only exhibited in cases of monstrous development, may in like manner have been choked by the compression of some contiguous parts, which got the start of them in the progress of their growth. It is equally easy to comprehend, that two contiguous parts may be constantly predisposed to graft together, long before we can trace them in a detached state. We perpetually see apples, peaches, and a variety of other fruits, become double, owing to the great facility with which their tissues graft together, when brought into close contact; and we can readily imagine that the tissues of two contiguous organs, whilst they are yet in their nascent state, must be in a condition even still better adapted for receiving this impression, than they would be at a later period of their growth.

In those cases of adhesion where the union is most perfect, it generally happens that some portions have necessarily become suppressed, and thus a monstrous form is produced, in which the number of its parts will lie between the regular number in a single flower, and some multiple of that number. Now, that which is so evidently the result of a natural grafting of contiguous parts, in these monstrous cases, may be conceived also to exist in other instances, where the same cause may have been in operation, previous to the very earliest stage of development to which the existence of the flower can be traced.

(119.) *Supernumerary Whorls.*—It sometimes happens, that a supernumerary development takes place, of one or more entire whorls, or of the parts of a whorl. In this way, certain flowers become double; but such are not necessarily barren, as is the case where double flowers have resulted from the transformation of the stamens and pistils into petals. The various parts of these supernumerary whorls alternate with those which precede them in the series.

(120.) *Normal Characters.*—It will readily be understood, how numerous may be the modifications which can be referred to the same normal condition of the parts of a flower, — if we suppose the three causes which we have enumerated, capable of acting separately, or together. If, for instance, the normal character of a flower consisted of five sepals, five petals, five stamens, and five carpels; and these several parts were so arranged, that all those which were in any one whorl, alternated in position with those in the contiguous whorls—this arrangement would constitute a highly regular flower, such as we meet with in the genus Crassula (*fig.* 134.). By simultaneously suppressing one, two, three, or four parts of each whorl, we may conceive four other flowers to be formed, equally symmetrical with the original, but disagreeing with this normal type, in not possessing a quinary arrangement of their parts. Irregularity might now be introduced, by suppressing certain parts of some whorls and not of others, or by forming adhesions between two or more parts of one whorl, whilst the other parts remained free; or by supposing some of the parts of one whorl to degenerate, and assume a variety of distorted shapes. In this way, an infinite variety of forms may be supposed to result from a few normal types; and it is by detecting these, that the systematic botanist is enabled to ascertain the affinities of certain species, which at first sight appear widely separated.

134

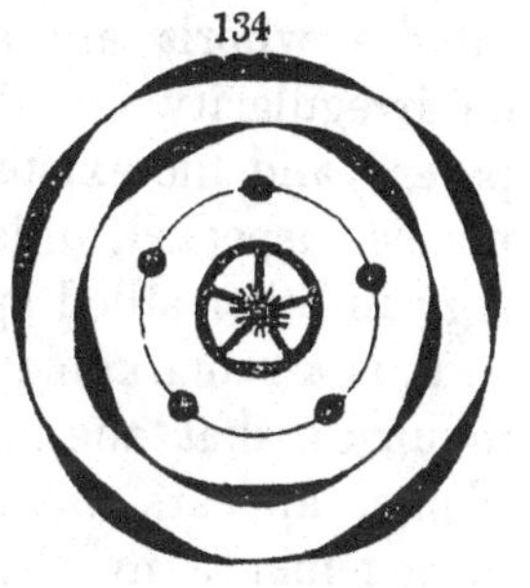

Whenever the parts of one whorl are placed opposite, instead of alternate with, the parts of the contiguous whorls, this circumstance is considered to indicate a want of regularity in the flower, although there may be no real want of symmetry in the arrangement; and such a state of things is always supposed to have originated in the abortion of one or more of the whorls. These whorls may possibly be still developed under certain conditions, and then the regularity of the flower would be restored, and the normal condition exhibited. In the annexed figure (135.) there are five whorls ascribed to the normal condition of certain organs, which "alternate" with each other in some flower; and by suppressing the parts in the second and fourth whorl, those in the first, third, and fifth are brought "opposite" to each other. Where two contiguous whorls are abortive, no irregularity would be apparent, and the existence of the suppressed parts might not be suspected, unless it were indicated by some analogy in other allied species.

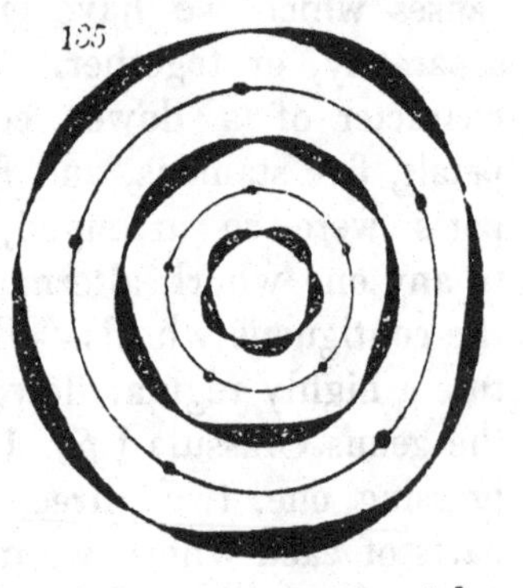

It is a remarkable circumstance connected with these inquiries, that the normal condition of dicotyledonous plants, appears most frequently to involve a quinary arrangement, in the disposition of the subordinate parts of the several whorls; whilst that of Monocotyledons, equally affects a ternary. In a multitude of examples, where the parts or organs of the class exceed these numbers respectively, they are still observed to be some multiples of them — 10, 15, 20, &c., or 6, 9, &c.; and many deviations from this rule, are clearly referable to the abortions of some of the parts, and the adhesions of others; so that a considerable approximation has apparently been made, towards the discovery of some general laws on this subject.

(121.) *Spiral Arrangement of foliaceous Appendages.*—The variety exhibited in the disposition of leaves, and other foliaceous appendages to the stem, or other axes, may be reduced to a general mode of expression, by a method proposed by M. Schimper, and subsequently elucidated by M. Braun. Even in those cases where their distribution does not seem to be regulated by any law of symmetry, this may be considered to be owing to the various disturbing causes which are perpetually modifying the conditions under which their arrangement would otherwise have taken place. As the mineralogist refers the crystalline forms of his minerals, to certain geometric solids, whose angles at least are the same as those on the crystal; so we must here neglect the accidental displacements, produced by the unequal development of those parts to which the foliaceous appendages are attached, or some other circumstances, and look only to the primary conditions upon which their distribution depends. If in those cases, even, where the leaves are most scattered on the plant, we were to draw a line from any one which is seated lower down the stem or branches, to another next above it, and so on, this line will be found to follow a spiral direction; and thus we ultimately arrive at a leaf, which is seated vertically above that from which we started. The usual mode of expressing this, is to name the number of the

136

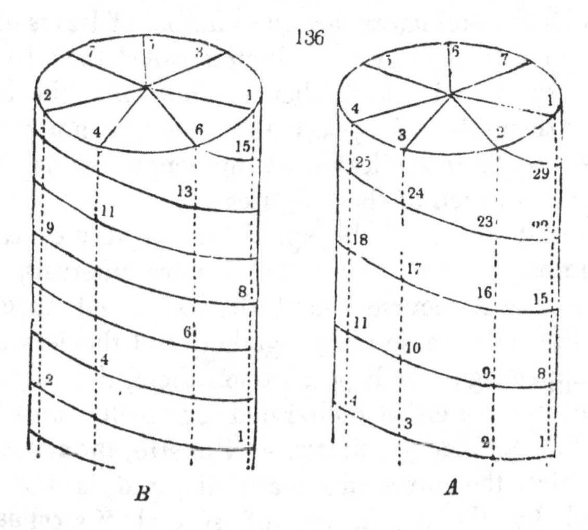

leaf which ranges vertically over the first on this spiral, but without any reference to the number of coils which the spiral makes before this happens. Thus, in each of the annexed figures (*fig.* 136.), No. 8. ranges vertically over No. 1.; but, in *A*, this happens after one coil; and in *B*, not until after three coils of the spiral. The numbers are ranged at equal intervals, indicated by the eight vertical lines drawn on the surface of the cylinder.

(122.) *Divergence of generating Spirals.* — M. Braun proposes to note the nature of this arrangement, by giving it a numerical value, which shall be expressive of the angular distance between two successive leaves on the spiral, when they are projected on a plane perpendicular to the axis. The expression obtained, is termed the "divergence" of the generating spiral. Thus, the divergence in *A*, is the angular distance between 1 and 2 (viz. $\frac{1}{7}$ of the circle); but the divergence in *B*, is $\frac{3}{7}$, as may be seen by inspecting the summits of the two figures. The numerators of these fractions also express the number of coils which the generating spirals make, before one leaf ranges vertically over another; and their denominators, are the number of leaves distributed upon this interval — which is called the "length" of the spiral. It is further evident, that the leaves arrange themselves along as many lines drawn parallel to the axis, as there are leaves on one length of the spiral, viz. seven in each of these figures.

Where the coils of the spiral are not very close, and the numbers succeed each other at short intervals, it is easy to trace its course round the axis; but, in many cases, the coils are so close together, and the leaves, or other appendages, so disposed upon them, that all traces of its course are either obliterated, or much confused.

(123.) *Secondary Spirals.* — But still, the symmetry with which the leaves are really disposed, is now manifested by the appearance of several "secondary" spirals, which may be traced in various directions. This is well exhibited in the arrangement of the scales of a fir-cone (*fig.* 137.); and we shall endeavour to show, how the real disposition of the scales on the "generating" spiral may be readily ascertained, from merely inspecting the appearances presented by these secondary spirals. Thus, in the spruce fir (*Pinus abies*), it is easy to trace several sets of spirals, running parallel to 1, 9, 17, 25, &c.; and other sets parallel to 1, 6, 11, 16, &c.; and others to 1, 4, 7, &c., and so on. In the present example, there are twenty-one lines which may be drawn through those scales which are ranged vertically over the others, as 1, 22, 43, &c., 14, 35, 56, &c. and so on. This number, as was before shown of the seven verticals, in *A* and *B* (*fig.* 136.), indicates the number of scales that are ranged upon one length of the spiral. But the course of the generating spiral is not apparent, and, consequently, the numerator of the fraction which expresses the divergence is unknown.

137

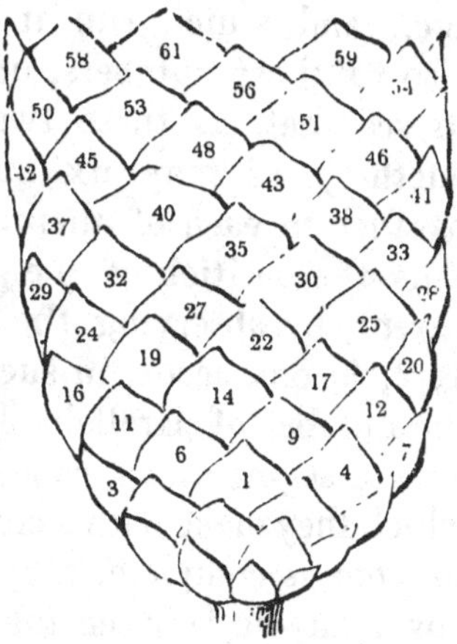

(124.) *To fix Numbers to the Scales.* — We may easily observe, that the numbers on the scales which form the different secondary spirals, are in arithmetical progression; and we shall presently show, in the next article, that the common differences in these progressions, also indicate the number of similar secondary spirals which range parallel to each other. Thus, there are *eight* parallel spirals, 1, 9, 17, &c., 6, 14, 22, &c., where the arithmetical progressions have all the same common difference — *eight*. Hence we see a ready means of numbering the scales on the cone, without the necessity of previously ascertaining the course of the generating spiral. Fixing on scale (1) for a beginning, and counting the number of parallel spirals (viz. eight) which run in one direction, as above, we can fix the numbers

1, 9, 17, &c. on one of these spirals; then counting the number (viz. five) which lie parallel with 1, 6, 11, &c., and which run in a contrary direction, we can also fix those numbers, upon that spiral: and it is easy to see that, as these two sets of spirals intersect one another, we may fix numbers to every other spiral parallel to each of them, that is, to every scale; and thus the position of the generating spiral becomes apparent, by observing the scales on which the numbers 1, 2, 3, &c. occur, in succession. We may easily count the number of parallel spirals of the same class, even in a mere segment of a cone, by observing the intersections which they make with a circle drawn round it; and, where the cone is complete, they may be counted, by observing how many lie between the coil which completes a length, in one of them. Thus the spiral 1, 6, 11, 16 . . . 38, 46, 51, 56, has four others lying parallel to it, and between two of its successive coils; there are, therefore, five such spirals in all, and, consequently, the common differences on them are five. Looking to the truncated edge, we might ascertain the same fact, by observing that five such spirals meet it in the scales 59, 61, 58, &c. Also eight parallel spirals meet it in the scales 54, 59, 56, 61, 58, &c. But even without numbering many of the scales, we may ascertain, first the denominator, and then the numerator, of the fraction which expresses the divergence. We need only place the numbers 1, 9, 17 in one direction, and then pass from 17 to 22 in another direction, and we arrive at the scale placed vertically over number 1; and thus we know that 21 is the denominator of the fraction. To find the numerator, we must fix the scales 2 and 23 — the latter ranging vertically over the former; and then fixing all the scales that lie between the verticals (1, 22,) and (2, 23), which we shall find to be 9, 17, 4, 12, 20, 7, 15 — through each of which other verticals may be drawn — we obtain the angular distance between any two vertical lines, viz. $\frac{1}{8}$ of a circle: and this gives the number 8, for the required numerator. This may perhaps be rendered more evident by an inspection of the annexed figure (138.), which shows the relative position of the scales on one length of the spiral, seen in the direction of the axis.

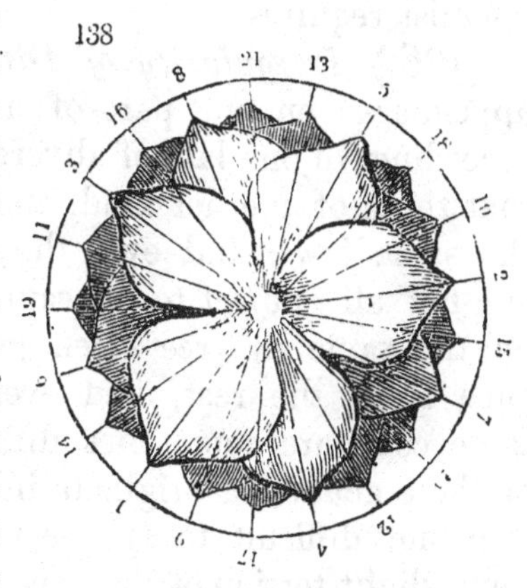

138

(125.) *Number of secondary Spirals.* — Although the number of secondary spirals which are readily distinguishable, is limited, yet it is evident that we may really establish the existence of any number, however great, by merely passing a line successively, from No. 1 to any other scale, and so on to that scale next beyond it, which has the same relative position towards it, as it has to No. 1. In other words, we may have arithmetical progressions with all possible common differences, which shall represent different secondary spirals; and these spirals may be coiled, some to the right, and others to the left. We proposed to show (what we took for granted in the last article) that the number of parallel spirals of the same class, was always equal to the common differences, of the progressions on these spirals. It is clear that the generating spiral, passing successively through 1, 2, 3, &c., must be unique: but the secondary spiral, which passes through the odd numbers, 1, 3, 5, &c., leaves the even numbers, 2, 4, 6, &c., which form a second spiral, of the same class; that is to say, there are two secondary spirals, where the common difference is 2. There are three spirals, in the same manner, which pass through 1, 4, 7, &c., 2, 5, 8, &c., 3, 6, 9, &c., where the common difference is 3; and so on of all the rest. Several other properties, of a mathematical nature might be mentioned; but sufficient has been said, to show the simplicity of the investigations necessary for obtaining an expression for the divergence, which is all that the botanist requires.

(126.) *Irregularity of Divergence.* — Although the appendages on one part of a plant, may be arranged according to one law of divergence, it does not follow that those of another kind, and on another part, possess the same law; and even the same kind of appendages are not all subject to the same law: thus, a few cones on the same fir tree often possess a different divergence from the rest, and even different parts of the same cone are sometimes differently disposed. Many of these anomalies originate in disturbing causes, which it is not difficult to appreciate; such, for instance, as some slight torsion of the axis, or the abortion of some of the parts, &c. It is also common to find the generating spiral turning to the right in some cones, and to the left in others, upon the same tree.

(127.) *Examples of Divergences.* — From what has been said, it will readily be seen, that the disposition of foliaceous appendages may be conveniently and accurately expressed, in terms of the divergence of the scales on the generating spiral, unless they happen to be so irregularly disposed as to lose all traces of a symmetrical arrangement. Thus, where the appendages range in a line along one side of the axis, the divergence is $= \frac{1}{1}$; where they are ranged in two rows, on opposite sides of the axis (*distichous, fig.* 139.), the divergence $= \frac{1}{2}$; when in three rows (*trifarious*), the divergence may be $\frac{1}{3}$ or $\frac{2}{3}$: the latter, however, may be considered the same as $\frac{1}{3}$, turning round the axis in an opposite direction. One of the most common, the "quincunxial" arrangement, where the appendages range in five ranks, may be produced by four different divergences, represented on the circles in the annexed figure (*fig.* 140.); but here also it will be seen, that two of them are the same as other two, only that the spirals turn in opposite directions. And always, where the denominator of the fraction is a prime number, there will exist one number less than that of the divergences, according to which the generating spiral may be constructed — and a similar number of vertical ranges will still be the result. But where the denominator is not a prime number, then some of the fractions which express these different divergences, are not in their lowest terms; and these divergences represent the very same spirals as when such fractions are so reduced. Thus, when there are six vertical ranges (*fig.* 141.), the divergences may be taken as $\frac{1}{6}$, $\frac{2}{6}$, $\frac{3}{6}$, $\frac{4}{6}$, $\frac{5}{6}$; but $\frac{2}{6}=\frac{1}{3}$, and $\frac{4}{6}=\frac{2}{3}$, both of which represent the trifarious arrangement; also $\frac{3}{6}=\frac{1}{2}$, which is the distichous. Hence $\frac{1}{6}$ and $\frac{5}{6}$ are the only divergences which represent the hexafarious arrangement, and even these may be reduced to one kind, only the spiral would be turned in opposite directions in the two cases.

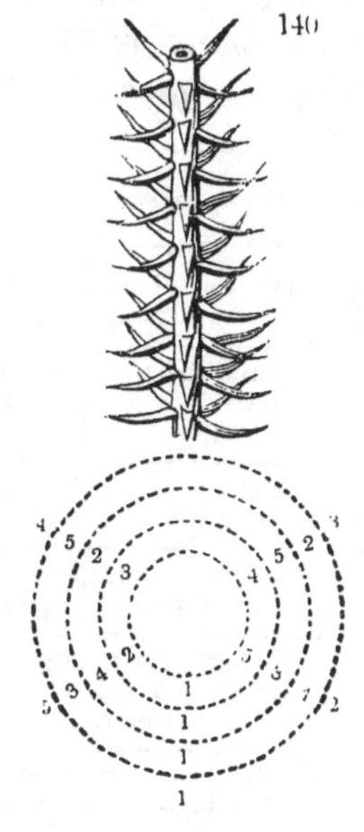

140

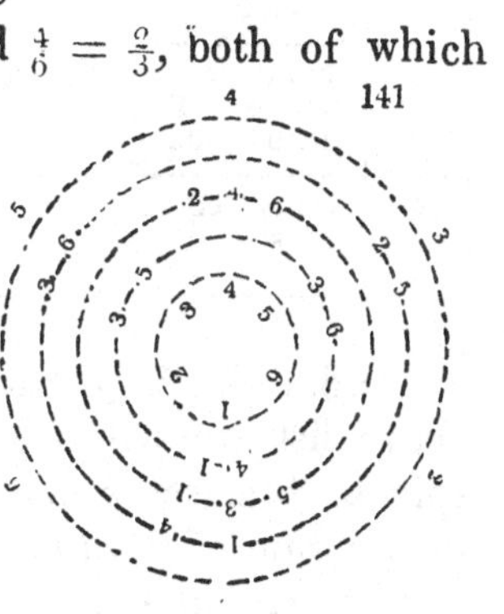

141

Examples of various Forms of Divergence among certain Species of the following Genera, selected from a long List given by M. Braun.

Div.	*Dicotyledones.*	*Monocotyledones.*	*Acotyledones.*
$\frac{1}{2}$	Asarum ; Tilia ; Vicia; Orobus.	Spikes of all Gramineæ ; Cyperus; Acoruscalamus.	Fissidens ; Didymodon capillaceus.
$\frac{1}{3}$	Cactus triangularis.	Carex ; Colchicum autumnale.	Gymnostomum æstivum; Jungermannia trichophylla.
$\frac{2}{5}$	Common in this class.	Scirpus acicularis ; Schœnus fuscus.	Common.
$\frac{3}{8}$	Laurus nobilis ; Ilex aquifolium.	Lilium candidum ; Scirpus lacustris.	Commonest in mosses; Lycopodium Selago.
$\frac{5}{13}$	Euphorbia segetalis ; Convolvulus tricolor.	Agave Americana ; many Orchis.	Orthotrichum affine ; Aspidium filix mas.
$\frac{8}{21}$	Isatis tinctoria ; Plantago lanceolata.	Orchis conopsea ; many Yuccæ.	Hypnum alopecurum ; Polytricum piliferum.
$\frac{3}{34}$	Euphorbia cæspitosa ; Plantago media.	Yucca aloefolia ; Ornithogalum pyrenaicum.	Sphagnum ; Politrichum formosum.
$\frac{2}{55}$	Cactus coronarius.		

(128.) *Mode of examining the Divergence.*—To the above list we will add a complicated example, in the spinous bracteæ which compose the involucrum of *Carduus Eriophorus*, and explain the manner in which the divergence may be ascertained. It is easy to observe two sets of spirals respectively parallel to *A B* and *C D* (*fig.* 142.), of which there are 34 of the former, and 21 of the latter. Fixing the Nos. 1, 35, 69, in one direction, and 90 in the other, as in art. 124., we arrive at the bractea which ranges vertically over No. 1.

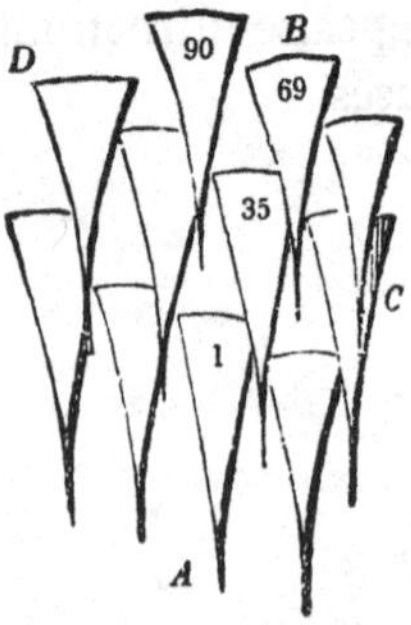

Also No. 35, is evidently nearer than any other bractea to the vertical line through 1 and 90. To construct the figure which represents the projection of one length of the generating spiral, we may thus proceed. Place No. 1 in the circumference of the circle (*fig.* 143.), and divide it into 89 equal parts; place No. 35 on the part nearest to No. 1: and 34 is the common difference on that secondary spiral, which is more nearly perpendicular than any of the others. The series on this spiral is, therefore, 1, 35, 69, 103, &c., of which we may place 69 on the next division to 35; but as 103 belongs to a second length of the generating spiral, we must subtract 89 from it, and thus we shall obtain No. 14, which ranges vertically below it, and is, consequently, within the first coil of the generating spiral itself, and therefore succeeds No. 69, on the circle. From No. 14 then, we may begin with another secondary spiral, whose common difference is the same as the last; and, consequently, we place the Nos. 48, 82, next in succession to 14; but 106 rises into the second length of the generating spiral, and we must subtract 89 as before, which gives us No. 17, for the next number in the circumference of the circle which represents only the first length. And so on until we arrive at No. 2. We shall thus ascertain that No. 2 is placed at 55 intervals from No. 1, and, consequently, that the divergence in this example is $= \frac{55}{89}$. It may readily be understood, by any person accustomed to mathematical investigations, that the first term common to the two arithmetical series, 1, 35, 69, &c., and 2, 91, 180, &c. (and which is 1871), will be the number on the bractea intersected by that spiral, which is represented by the first of these

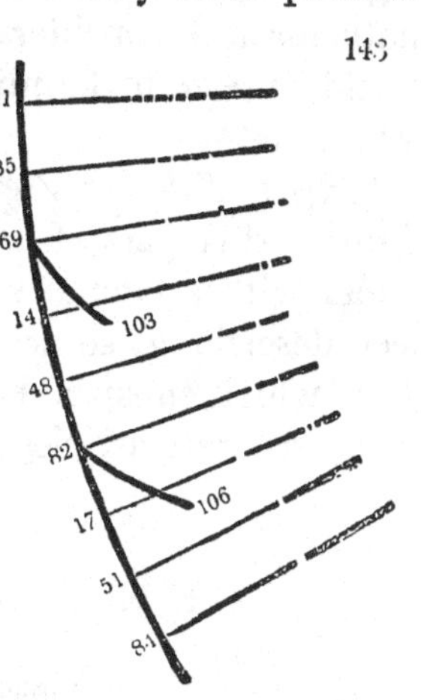

arithmetical series, and the vertical line through No. 2, represented by the second; and also that one less than the number of terms in the first series represents the angular distance of 2 from 1. Several other interesting mathematical considerations might be given, but they would appear to be misplaced in a treatise of this description.

(129.) *Tabular View of Vegetable Organs.*—In concluding this part of our subject, we shall present the reader with a tabular view of the various organs we have been describing, so arranged as to display the subordination which subsists between them; giving a reference to the separate articles in which each is described.

I. Elementary Organs (13.).

		Modifications.
Membrane (13.) Fibre (13.)	Vesicles (16.) Tracheæ (23.) Ducts (24.) Vital vessels (27.)	Cellular tissue (16.) Vascular tissue (22.)

II. Compound Organs (28.).

Pellicle (29.) Stomata (30.)	Epidermis (29.)
	Hair (31.) Stings (31.) Glands (31.)

III. Complex Organs (32.).

* *Nutritive* (38.).

Spongioles (39.) Fibrils (39.)	Roots (39.)	
	Appendages (41.)	
Pith (48.) Medullary sheath (49.) Woody layers (50.) Alburnum (50.) Medullary rays (51.) Liber (52.) Cortical layers (52.)	Stems (44.) and Branches (59.)	Thorns (62.) Bulbs (65.) Tubers (64.) Suckers (62.) Runners (62.)
Petiole (69.) Limb (69.)	Leaf (69.)	Phyllodia (75.) Spines (78.) Tendrils (79.) Pitchers (80.)
	Stipules (77.)	

** *Reproductive* (85.).

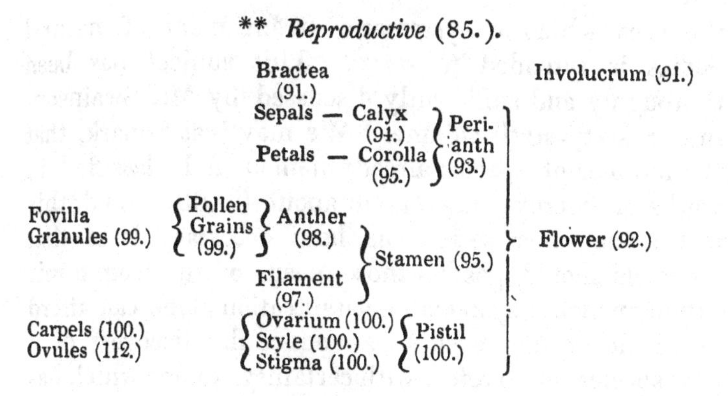

COMPOSITION OF THE RIPE FRUIT (105.).

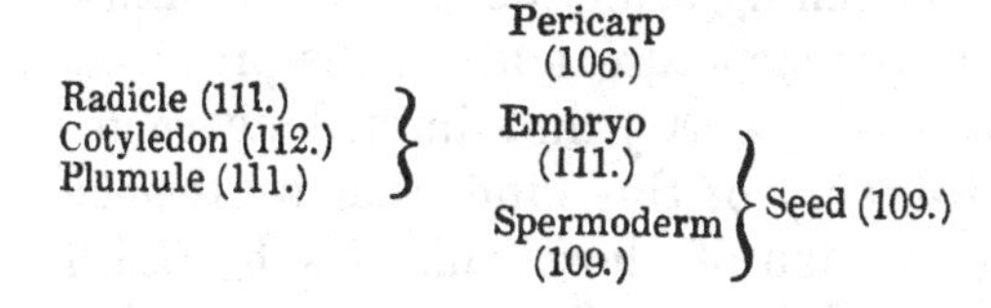

SECTION II.

TAXONOMY AND PHYTOGRAPHY.

CHAP. VII.

NATURAL GROUPS (131.). — VALUES OF CHARACTERS (132.) — SUBORDINATION OF CHARACTERS (133.). — NATURAL ORDERS (135.). — ARTIFICIAL ARRANGEMENTS (136.). — LINNÆAN SYSTEM (137.). — APPLICATION OF IT (140.).

(130.) *Taxonomy.* — WE have no space to devote to any extended review of the various methods and systems which have been proposed for the classification of plants; and it is not necessary for us to explain the uses which a systematic arrangement of natural bodies is intended to serve. This subject has been thoroughly and sufficiently discussed by Mr. Swainson, in our sixty-sixth volume. We may just remark, that the number of species already named and classified in works of botany, amounts to about 60,000; and this fact alone must satisfy us, how necessary it is that botanists should possess those means of intercommunication, which a systematic classification alone can afford — whenever they wish to announce the discovery of a new species, or to refer, with certainty, to one which has been previously noticed. But, if we have the higher object in view, of searching after the laws and principles which regulate the structure and fix the properties of plants, then it is a necessary and immediate consequence of every discovery of this kind, that we thereby obtain a nearer conception of those affinities by which plants approach, and of those differences by which they recede from each other; and this, in fact, amounts to a closer insight into that hitherto undiscovered system, or plan, upon which we must feel satisfied that the Author of nature has proceeded in creating all natural objects.

(131.) *Natural Groups.* — We have already (art. 33.) mentioned the leading characteristics of the three primary groups, or classes, into which plants seem to be naturally divisible. Each of these, again, admits of subdivision into minor groups, which severally contain such species as are more nearly related to each other than to those of other groups. By further subdivisions of this kind, a subordination of groups, of smaller and smaller dimensions, is obtained, until we arrive at those groups which do not readily admit of further subdivision, and which are termed "genera." It must, however, be obvious that this method, of analysis, is not the actual process in which the primary groups were originally established. This was effected by a synthetical mode of procedure — by comparing separate individuals, and by selecting those which most nearly resembled each other; and thence establishing, in the first place, the limits within which a given species might be supposed to vary. Then, by comparing different species, and selecting those which had the greatest resemblance, a genus was constructed. Then the genera were grouped into orders; and lastly, those orders which possessed only a few general but important points of resemblance, were arranged under the three classes alluded to. But when these several groups were once established, a further refinement in their classification could be made; and the principles upon which this was effected, may be explained by the analytical process to which we have just had recourse, when we said that all species are comprised, first, in a class; secondly, in an order, or family; and thirdly, in a genus. In very many cases, a further subordination may be established among the several groups; and, from various considerations, they may either be aggregated into larger, or subdivided into smaller groups; to which other names are applied, of which we have given an example in art. 102. When any group is subdivided into larger groups than those which it is supposed to contain under the system of subordination already described, these are generally recognised by the addition of the word "sub" to the name of the original group; thus we have sub-classes, sub-orders, and sub-genera. Certain groups are also termed "Tribes," "Cohorts," "Sections," and "Divisions;" and some of these terms are used indiscriminately for subordinate groups among the classes, genera, and even species. When a "variety" of any species is reproducible by seed, and retains its peculiarities pretty steadily, without returning to the more common type, it is termed a "race;" but when its distinguishing characters are transient, and may be modified by a change of soil or situation, it is only a "variation." In this way then, we establish a subordination among the natural groups into which plants may be arranged, and which may be exemplified by the following instance.

I.	Class - - - -	Dicotyledones.
*	Sub-class - - -	Calyciflorae.
II.	Order - -	Leguminosæ.
*	Sub-order - - -	Papilionaceæ.
**	Tribe - - -	Loteæ.
***	Sub-tribe - -	Genisteæ.
III.	Genus - - -	Anthyllis.
*	Sub-genus (or Section) - -	Vulneraria.
IV.	Species - - -	Vulneraria.
*	Variety - - -	Dillenii.
**	Race - - -	Floribus coccineis.
***	Variation - - -	Foliis hirsutissimis.

(132.) *Value of Characters.* — In determining the particular group to which a plant belongs, it is necessary to compare its "characters" with those of other species. By the term "characters," we mean the peculiar appearances presented by different organs. Thus, a leaf may be round, lanceolate, &c.; the petals may be united, abortive, &c.; and these adjectives denote the peculiar characters of these organs. It will readily be understood, that some characters must be of much greater importance than others, in determining the affinities of different species. Thus, the first degree of affinity in phænogamous plants, is almost always to be ascertained by a single character, residing in the embryo; and we may determine at once, to which of the two primary groups it belongs, by attending to this circumstance alone. But even here, this primary character may be so far disguised or modified, as inevitably, in some instances, to lead us into error, if it were not possible for us to check our observations by other considerations, of secondary importance in most cases, but which, in the present instance, are quite sufficient to correct our judgment, and to satisfy us of the real affinities of the plant in question. Thus, in the genus Cuscuta, the character of the flower, the structure of the stem, and other circumstances, clearly indicate that it belongs to the class "Dicotyledones" — although the embryo has no cotyledons, and the stem is leafless. The inference to be drawn from these facts is, that the cotyledons and leaves are abortive; and hence we might expect, if ever such a phenomenon should occur as a leafy Cuscuta, that its cotyledons would certainly resemble those of other Dicotyledones. When the class of any plant has been determined by the presence of some one character, or by the combination of several, we next renew our search for other characters of a less general description, to ascertain the "order" to which it belongs. And when we have found the order, we must descend to still more minute particulars for fixing the "genus." It is, therefore, of the utmost consequence to these inquiries, that an accurate subordination of characters should be established; and for this purpose a few rules have been framed, which are the result of an extended examination of facts, or the deductions of common sense. We must remark, that a direct comparison can only be made between two organs which belong to the same class of functions: the nutritive organs must therefore be compared together, and the reproductive together, in order to establish a subordination in each series respectively. We may, however, afterwards determine, whether one of these two functions can not be considered more important than the other; and then we shall also be able to establish something like a fresh relation, between the several degrees which had been previously settled for the two series of organs. Suppose, for example, it were determined, that the cotyledons are among the organs of most importance to the nutritive system, and the root among those of the next degree. Suppose, also, the stamens were determined to be organs of the highest importance to the reproductive function, and the corolla among those of the next. Now, if it were also determined that the nutritive function was of more importance than the reproductive, then the cotyledons will be of more value than the stamens. But, although the root may be of more importance than the corolla, it does not follow that it is necessarily of more than the stamens; it may be of equal or less importance. In this latter case, we are comparing an organ of second-rate importance in the one series, with one of first-rate in the other.

If, we could determine the natural affinities of all plants, from a comparison of the characters deduced from one series alone, and could likewise determine their natural affinities from characters belonging to the other series, it is evident that the two arrangements thus established would strictly coincide. In the establishment of the minor groups, botanists have recourse almost exclusively to the reproductive organs; as their characters are much better defined, and more varied than those of the nutritive organs. The larger groups, however, are chiefly determined by characters belonging to the nutritive and elementary organs, as we have shown (art. 33.), where the exogenous structure tallies with the dicotyledonous embryo, and the endogenous with the monocotyledonous.

(133.) *Rules for fixing Subordination of Characters.* — The following rules may be advantageously consulted, for determining a subordination of characters in one or the other series.

1. Where two organs, belonging to different classes of functions, have the same relative value in their respective series, that organ will possess the greatest value which belongs to the most important function.

2. Those organs of the same series, are of the greatest value, which are of most general occurrence. Thus the cellular tissue, which is universally present, is the most important element in vegetation.

3. The adhesion which frequently subsists between an inferior and a superior organ, serves to point out the relative value of any two of the former; since it will be the same as that which was previously established for those of the latter, to which they respectively adhere.

4. The greater degree to which an organ is liable to vary, indicates an inferiority in its value. Thus the shape of the leaves, is of little importance beyond determining the specific distinctions of plants, and in

many cases is even of no further use, than in discriminating certain varieties of the same species.

5. The relative periods at which different organs are formed and developed, may also be taken as some test of their relative importance; those which are the earliest formed, being considered more important than others with which they are immediately connected, and of the same class.

By attention to these and a few other rules of less general application, a subordination of characters has been established, of which the chief results are exhibited in the following table: —

Relative Values.	*Elementary.*	*Nutritive.*	*Reproductive.*
1.	Cellular Tissue	—	—
2.	Vascular Tissue (*a*) Tracheæ (*b*) Ducts Stomata	Embryo and Sporule (*a*) Cotyledons (*b*) Radicle (*c*) Plumule	– –
3.	—	Root, Stem, Leaf, Frond, Thallus	(1) Stamens and Pistils. (2) Fruit, Pericarp, Theca.
4.	—	—	Perianth. (*a*) Corolla. (*b*) Calyx.
5.	—	—	Inflorescence, Torus, Nectary, Bractea, Involucre.

(134.) *Relative Importance of similar Organs.*—Besides the relative values of different organs, established in this table, we may estimate the relative value which two organs of the same kind bear to each other, in different species. This will depend upon the greater or less perfection which they exhibit in their respective modes of development; also, upon their position, connection with other organs, and numerous other particulars which it is impossible to define with any degree of precision, and which practice alone can enable the systematic botanist duly to appreciate.

(135.) *Natural Orders.*—As we make no pretensions in this volume to enter upon the details of systematic botany, we do not consider it advisable to present the reader with a bare enumeration of the characters of the natural orders which have been hitherto established in the most recent works. We shall content ourselves with explaining the connection which subsists, between the principal groups under which Jussieu arranged the natural orders, so far as they had been established in his time, with the principal groups in the recent system of De Candolle, under which this eminent botanist has arranged the natural orders as they are at present understood. Jussieu threw the natural orders or families with which he was acquainted, into fifteen groups, which he termed classes, and these he further combined into six principal groups or divisions; of which four belonged to Dicotyledones, and one each to Monocotyledones and Acotyledones. De Candolle has also four groups for the Dicotyledones and one for the Monocotyledones, but somewhat differently arranged; and he has split up the Acotylodones into two parts, one of which (although cryptogamic like the other) he classes with the Monocotyledones, and retains the other only as Acotyledones. He further arranges the whole of vegetation under two principal heads, according as plants possess, or are entirely without, any portion of a vascular structure.

Comparative View of the Systems of De Candolle and Jussieu.

Primary Divisions of De Candolle.	*Subordinate Groups (Classes of Jussieu) common to both.*	*Primary Divisions of Jussieu.*
* Vasculares *seu* Cotyledoneæ.		
A. Dicotyledoneæ *seu* Exogenæ.		C. Dicotyledones.
I. Thalamifloræ.	14. Hypopetalæ	III. Polypetalæ.
II. Calycifloræ.	13. Peripetalæ	
	12. Epipetalæ	
	11. Epicorollæ corisantheræ	II. Monopetalæ.
	10. Epicorollæ synantheræ	
	9. Pericorollæ	
III. Corollifloræ.	8. Hypocorollæ	
IV. Monochlamydeæ	7. Hypostamineæ	I. Apetalæ.
	6. Peristamineæ	
	5. Epistamineæ	
	15. Diclines	IV. Diclines.
	* Angiospermæ	
	** Gymnospermæ	
B. Monocotyledoneæ *seu* Endogenæ		B. Monocotyledones.
V. Phanerogamæ	4. Monoepigynæ	
	3. Monoperigynæ	
	2. Monohypogynæ	
VI. Cryptogamæ		A. Acotyledones.
** Cellulares *seu* Acotyledoneæ	1. Acotyledones	
VII. Cellulares		

We have explained in art. 102. the meaning of the terms which designate the principal groups of De Candolle in the first column of this table; and we shall now explain those which have been proposed for the classes of Jussieu, in the second column, as their etymology may assist the reader in recollecting them. They are combinations of words expressive of the three modes of floral arrangement described in art. 101., applied respectively (in the Dicotyledones) to the "petals," when these organs do not cohere together; to the "corollæ," when they are monopetalous; and to the "stamens," when the perianth is single. Thus, Epicorollæ indicates, that a monopetalous corolla is epigynous in the 10th and 11th classes; which are further distinguished from each other by the anther being united together (συν) in the 10th, and separate (χωρις) in the 11th. The term Diclines indicates the flowers of the 15th class to be unisexual; and in the two subdivisions of this class, the seeds are contained in a pericarp or distinct vessel (αγγος) in the one, and are without it, or naked (γυμνος), in the other. The derivation of the classes of the Monocotyledones is evident.

(136.) *Artificial Arrangements.*—An artificial arrangement proceeds upon the fact, that certain organs, in nearly all the species included under the same genus, have a great degree of constancy as to their number, relative size, position, and other characters; and these organs are selected as the basis of the systematic arrangement. Thus, for example, every species of the genus Ranunculus has more than twenty stamens, and these organs are similarly circumstanced with respect to the other floral whorls. The species of the genus Papaver, have their stamens arranged like those of the last-mentioned genus, and they are also numerous. These two genera belong to different natural orders, but they and many others are thrown together into the same artificial class, characterised by the species having their stamens numerous, and not attached to the calyx, the flowers also containing both stamens and pistils.

The natural groups, then, which we term genera, and which are the lowest in the rank of subordination, are not subdivided to suit the purposes of an artificial arrangement; but it is the higher groups only which are so. There are certain cases, however, where it is advisable to break through this rule, and to retain under the same artificial class, several genera of a natural order, which do not agree with the rule laid down for fixing their position in the system. In other words, it would be too great a violation of the natural group to which such genera belong, to separate them from it. Thus, for example, the greater number of those genera of the natural order Leguminosæ which have papilionaceous flowers, forming the tribe Papilionaceæ, have their filaments united round the pistil, so that nine are blended together, and one stands by itself (see art. 97.); and an artificial class (*Diadelphia*) has been constructed to admit all flowers which have their stamens united into two bundles. Now, there are a few genera of the Papilionaceæ, where the union of the ten filaments is complete; and these therefore strictly belong to another artificial class (*Monadelphia*), characterised by this circumstance. But in this case the natural affinity is so striking, that the artificial arrangement is broken through, and they are all classed together. We shall presently explain how the difficulty of such a false position is, to a certain extent, obviated. (Art. 138. *bis.*)

An artificial system which should disregard the construction of genera, and group species according to the principles of that system, would be the most perfect; but this would be descending to a degree of precision unnecessary for obtaining the sole purpose for which an artificial system should be employed, viz. the detection of the name of a plant; and the devices adopted for referring the anomalous species to their proper genus, and the anomalous genera to their proper class, are sufficient to counteract the smaller inconvenience of establishing a system at variance with these few cases.

(137.) *Linnæan System.*— The most celebrated of the several artificial systems which have been proposed, is that which Linnæus established, from considerations deduced from the number and disposition of the stamens and pistils; these organs maintaining a greater general resemblance in all the species of the same genus, and through many genera of the same natural group, than any others. They are at the same time sufficiently modified in different groups, to allow of these being thrown into several orders and classes, characterised by some definite and striking peculiarity. This system has been styled the sexual system. In his arrangement, Linnæus established twenty-four classes; the last of which embraces the whole of the natural class of Acotyledones, or flowerless plants.

The Dicotyledones and Monocotyledones are distributed unequally throughout the other twenty-three classes; some of these consisting entirely, or chiefly, of the one, and others of the other, whilst several of them are made up from both of these natural classes. The fundamental principles upon which his arrangement proceeds, are of the simplest possible description, but in the practical application of them, the beginner will unfortunately meet with several anomalies, and without repeated caution he is sure to be misled. The following table exhibits the names of the classes and orders of the Linnæan system; and we shall explain their etymology, as this is intended to convey the leading characteristic upon which each depends.

Tabular View of the Classes and Orders of the Linnæan System.

	Classes.		*Orders.*
1.	Monandria.	μονος.	Monogynia.
2.	Diandria.	δις.	Digynia.
3.	Triandria.	τρεις.	Trigynia.
4.	Tetrandria.	τετρας.	Tetragynia.
5.	Pentandria.	πεντε.	Pentagynia.
6.	Hexandria.	εξ.	Hexagynia.
7.	Heptandria.	επτα.	Heptagynia.
8.	Octandria.	οκτω.	Octogynia.
9.	Enneandria.	εννεα.	Enneagynia.
10.	Decandria.	δεκα.	Decagynia.
11.	Dodecandria.	δωδεκα.	Dodecagynia.
12.	Icosandria.	εικοσι.	Polygynia.
13.	Polyandria.	πολυς.	
14.	Didynamia.		Gymnospermia. Angiospermia.
15.	Tetradynamia.		Siliculosa. Siliquosa.
16.	Monadelphia.		Triandria, &c. as in the Classes.
17.	Diadelphia.		
18.	Polyadelphia.		

Class	Orders
19. Syngenesia.	Polygamia æqualis. superflua. frustranea. necessaria. segregata. *Monogamia.*
20. Gynandria. 21. Monœcia. 22. Diœcia.	Monandria, &c. as in the Classes.
23. Polygamia.	Monœcia, Diœcia, Triœcia.
24. Cryptogamia.	Filices, Musci, &c.

(138.) *Linnæan Classes.* — The first eleven classes are characterised by the "number" merely, of the stamens, which the species (or nearly all of them) in the respective genera contain; and their names are a compound of two Greek words, one of which signifies that number, and the other is *ανηϱ* (a man). The number eleven is not employed, as no flowers are found to possess that number of stamens. In the first ten classes, the species are pretty constant in the number of stamens by which their class is designated; but in the eleventh class the number is not so certainly fixed. There are, however, very few species included in it; and when the genera to which they belong have been once pointed out, the student is not afterwards likely to refer them to another class.

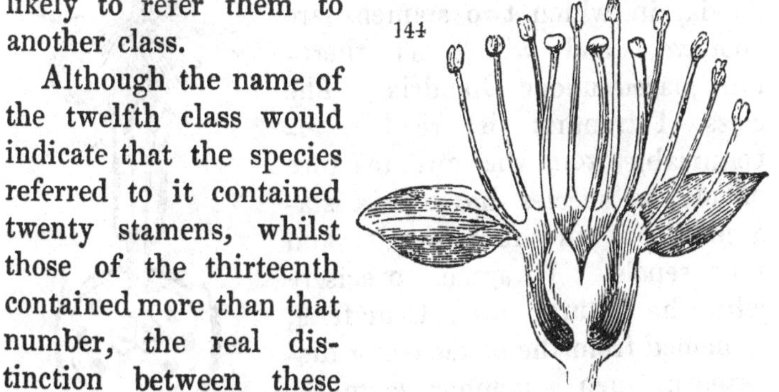

Although the name of the twelfth class would indicate that the species referred to it contained twenty stamens, whilst those of the thirteenth contained more than that number, the real distinction between these two classes depends more upon the position, than upon the number of these organs. In both classes the stamens are numerous — that is to say, are above a dozen in number; but in Icosandria they adhere to the calyx (*fig.* 144.), or are perigynous (see art. 101.); whilst in Polyandria they are free from the calyx, or are hypogynous.

The fourteenth and fifteenth classes are characterised by a twofold consideration, — the number and relative lengths of the stamens. In Didynamia there are four, and in Tetradynamia there are six; but the former is distinguished from Tetrandria, by two of the stamens being always shorter than the other two; and he latter from Hexandria by two being shorter than the other four. This is expressed by the word δυναμις (power), signifying that some of the stamens have an ascendancy over others, and this is combined with the word which expresses their number. These circumstances are not always readily recognised by beginners; and they should take into consideration a few other particulars which may enable them to correct their judgment. Thus, in Didynamia, the four stamens are not symmetrically disposed round the axis, but are thrown together on one side of the flower, which is always monopetalous, and never strictly regular. The lipped flowers (*Labiatæ*, art. 95. and *fig.* 93.) form a large portion of this class, excepting a few of them, as the genus Salvia, in which two stamens are abortive, and which is therefore placed under Diandria. The class Tetrandria is readily recognisable, from the circumstance of all its species having six stamens, but only four petals, and four sepals. It agrees precisely with the natural order Cruciferæ, so named from the petals being disposed in such a manner as to represent a cross (*fig.* 145. *a*). (*b*) shows the relative position of the floral organs.

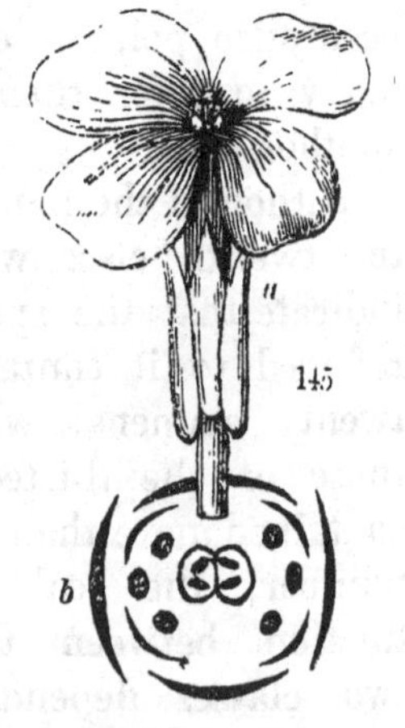

The names of the three next classes indicate that the filaments are united into bundles, expressed by the word αδελφος (a brother); these bundles or brotherhoods of stamens, being either one, two, or more than two respectively. Where there is only one (in Monadelphia), the filaments must necessarily form a cylindrical tube round the pistil (*fig.* 97. *a*). The greater portion of Diadelphia is composed of a large section of a natural tribe, the Papilionaceæ, belonging to the natural order Leguminosæ. (See art. 136.) A small section of the Papilionaceæ, in which the filaments are perfectly free from any adhesion, is classed under Decandria, in the same way as a few of the Labiatæ are placed under Diandria. The remainder of this artificial class is almost entirely composed of the few genera which belong to the Fumariaceæ and the Polygaleæ; the former having six, and the latter eight stamens, united into two bundles.

The class Polyadelphia is exceedingly small, (the genus Hypericum forming its most prominent feature,) and the stamens are here placed in little tufts or bundles round the pistil.

The nineteenth class is also strictly natural, like the fifteenth, coinciding with the natural order Compositæ, so named from the inflorescence being composed of a dense mass of small flowers, or florets (as they are here termed), closely invested by an involucrum. The whole head, in popular language is called a single flower. (See *fig.* 87.) The name of the artificial class signifies that the anthers are united, συν (together,) and γενεσις (generation).

Although the several parts of the florets are very minute, and the adhesion of the anthers into a tube round the style not readily recognisable, yet there is very little difficulty in referring any species of this class to its right position. There are a few flowers in some other natural orders, arranged in heads resembling those of the Compositæ, but their anthers are free.

The twentieth class is named from γυνη (a woman), and ανηρ (a man); the centre of the flower not having the pistils and stamens separate in distinct whorls, but grafted together into one column, on the summit of which the anthers are seated near the stigma. This class is principally made up of the natural order Orchideæ, which includes all those singular flowers commonly known by the name of orchises and air-plants.

The next two classes are characterised by having unisexual flowers, expressed by the word οικος (a house); intimating that, in Monœcia, flowers of both sexes are found on the same plant; whilst in Diœcia the stameniferous flowers are on one plant, and the pistiliferous on another.

In Polygamia, γαμος (marriage), we have three kinds of flowers, which may all, or some only, be placed on the same plant. In these cases, it should seem that the flower in its most perfect form contains both stamens and pistils; and that in those flowers, where either of these organs is wanting, it is from abortion, and not that any difference of construction precludes its development.

And lastly, Cryptogamia, from κρυπτος (hidden), and γαμος (marriage), there being no flowers apparent from whence seeds are produced.

(139.) *Linnæan Orders.* — The orders of the several classes depend upon circumstances, connected either with the stamens or pistils.

In the thirteen first classes, the orders are fixed entirely by the number of the pistils, and this is expressed by the word γυνη (a woman) in composition with the Greek words signifying the number present. In some compound pistils, however, this number is calculated from the number of the styles or stigmas rising from the top of the ovarium, when those organs happen to be remarkably distinct.

In class fourteen, there are two orders, characterised by the manner in which the ovaria are developed into seed-vessels. One (Angiospermia) is named from αγγος (a vessel) and σπερμα (a seed), and in this case the pericarp is composed of two carpels blended together into a single two-celled capsule, containing many seeds attached to a central placenta. The other order (Gymnospermia) was so named from a mistaken opinion that the seeds were destitute of any pericarp, or naked (γυμνος). In this order the pistil is composed of four carpels, each containing a single seed, and agglutinated together into a compound ovarium with one style. As the fruit ripens, the carpels separate, and ultimately become four nuts, seated at the bottom of the calyx. The two orders are, therefore, readily distinguished, by the former containing only one seed-vessel with many seeds, and the latter four seed-vessels which resemble four naked seeds.

The fourteenth class also contains only two orders, which are characterised by the comparative lengths of the seed-vessels. They are composed of two carpels united by their edges, and are divided into two cells by a transverse membranous partition (see art. 109. *fig.* 123.). When the length of the seed-vessel exceeds its breadth three or four times, it is termed a siliqua, and the order to which it belongs is named "Siliquosa." When the length and breadth of the seed-vessel are nearly the same, the order is named "Siliculosa." These distinctions are apparent in the flower, from the earliest stages of the ovarium, and long before it becomes a true seed-vessel.

In the sixteenth, seventeenth, and eighteenth classes, the orders depend upon the number of the stamens; and in this respect they resemble the thirteen first classes themselves.

The nineteenth class was originally divided into six orders; in five of which the flowers were aggregated into heads, and thence distinguished under the name of "Polygamia;" whilst the sixth contained those simple flowers, whose anthers, as in the violets (Violæ), were more or less united. But this last order has been abolished by the universal consent of botanists; and the species which it contained, are now referred to their position in the system, without regard to the syngenesious character of their anthers. Of the five orders, then, which it now possesses, the first, "Æqualis," is so named from all the florets being "alike;" each containing both stamens and a pistil (*fig.* 146. *a*). In "Superflua," the outer florets have a pistil but no stamens; whilst the florets in the centre contain both (*b*). In this case, the outer florets, as in the daisy, are "ligulate,"

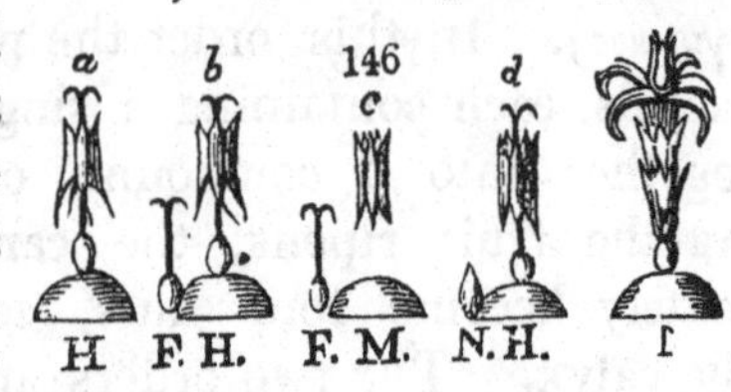

or "strap-shaped," and constitute what is termed the "ray;" whilst the inner florets are all "tubular," or "floscular," and form the "disk" of the capitulum. The inner florets being the most perfect, and sufficient to secure the production of seed, the outer florets appear as it were "superfluous," from whence the name has been given to the order. In "Necessaria," (*c*) the outer florets contain pistils only; and the inner, stamens only; and consequently both are "necessary" for perfecting the seed. In "Frustranea," (*d*) the central florets are perfect, or contain both stamens and a pistil; whilst those in the ray contain neither, and hence appear to be formed, as it were, in "vain" (frustra), as regards the perfecting of seed. The corolla of the latter florets is generally very highly developed, and assumes a handsome appearance, as in the genus "Centaurea" (*fig.* 87.). In "Segregata" (I), each floret is surrounded with a distinct and well-defined involucrum of its own, which "separates" it completely from the other florets in the same capitulum. In the diagram (*fig.* 146.), these different arrangements of the pistils and stamens are represented, and the capital letters further refer to the kind of florets of which the capitula are composed, viz. H (hermaphrodite), M (male), F (female), N (neuter), I (involucrate).

In the two next classes, Monœcia and Diœcia, the orders depend upon the number and arrangement of the stamens, precisely as in the several classes already enumerated; whilst in Polygamia the orders are characterised by the flowers being monœcious, diœcious, or triœcious.

There is no connection between the nomenclature of the orders of the class Cryptogamia, and the characters of the plants they contain; but some of them are familiar to most persons, as the ferns (Filices), mosses (Musci), seaweeds (Algæ), mushrooms (Fungi).

(138. *bis.*) *Application of the Linnæan System.*—Notwithstanding the apparent great simplicity of this system, there are many anomalous cases to which it cannot be directly applied. In order to meet these, Linnæus made use of an expedient by which such species as do not strictly belong to the class and order under which their genus is arranged, may still be ascertained. Their names are placed in Italics at the end of the order to which they really belong, and in which they would naturally be sought for; so that the student, who has not been able to detect them among the genera there enumerated, may refer to the index, and search among these anomalous cases. Thus, for example, the genus Gentiana is classed under Pentandria Digynia; but *Gentiana campestris* has generally only four stamens, and would be sought for under Tetrandria Digynia. Not being found among the genera there enumerated, it must be one of the few anomalous species, whose names are mentioned; and these must be all referred to, before it can be determined which of them it may be. The very unequal distribution of the classes is another inconvenience in this system. The great bulk of plants are included in about one half of them, whilst the others contain comparatively few. If, however, attention be paid to the general form of the flowers, the relationship which usually subsists between the divisions of the perianth and the number of the stamens, in such as have a regular corolla, and a few other particulars, the knowledge of which a little practice alone can bestow, these difficulties are soon greatly diminished, and many large na-

tural groups will be instantly referred to their proper class and order, without the necessity of searching for the characters upon which their arrangement depends. It will be soon seen that Triandria, Hexandria, and Gynandria contain the great bulk of the Monocotyledones, and that there are very few of this natural class among the other artificial classes. This circumstance is connected with the ternary arrangement of the subordinate parts of the floral whorls, to which we have alluded (art. 120.). On the other hand, the great bulk of Dicotyledons are included in those classes where some trace or other of a quinary disposition is observable. Thus, Pentandria, Decandria, Icosandria, and Polyandria are large classes answering to this description; and Syngenesia, which is the largest of any, has always five stamens, and the corollæ generally exhibit a tendency to a subdivision into five separate petals, indicated by five teeth at the end of the florets. Didynamia is eminently irregular; but even here, the normal character of the species seems to repose upon a quinary arrangement, which is sometimes manifested by a monstrous development of the suppressed organs, as in the varieties termed "Peloria," of the genera Antirrhinum and Linaria (see art. 114.). Tetradynamia is not unsymmetrical, but equally irregular, as regards the more usual characteristic of a dicotyledonous flower.

PART II.

PHYSIOLOGICAL BOTANY.

CHAPTER I.

VITAL PROPERTIES AND STIMULANTS.

VEGETABLE LIFE (139.). — PROPERTIES OF TISSUES (141.). — ENDOSMOSE (144.). — VITAL PROPERTIES (145.). — STIMULANTS TO VEGETATION (152.).

(139. *bis.*) *Vegetable Life* — HITHERTO we have been occupied with the forms only which the various organs of plants assume, and the manner in which they may be considered to be mutually related. We have been examining merely some of the details of that exquisite mechanism by means of which the vital principle is enabled to act and may be acted upon; and thus produce all the varied and complicated results which the phenomena of vegetation present. In this second part of our treatise, we propose to examine the vegetable machinery in a state of action, and to search for indications of those laws by which vegetable life enables the organic bodies to which it is united to grow and multiply. It would be an unnecessary waste of words to offer any proof that plants are organised bodies endowed with life. No one is so little observant, as to be ignorant of the more general phenomena of vegetation, that plants originate from seed, that they are gradually developed, and,

after having attained perfection, that they as gradually decay, die, and are decomposed. In fact the general phenomena of life and death, are scarcely less striking in the vegetable than in the animal kingdom; and probably the vital principle, considered apart from sensibility, is something of the same kind, if not the very same thing, both in animals and vegetables. This similarity or unity in essence must lead us to expect, what experience has shown to be the fact, that a considerable analogy exists between the functions of animal and vegetable life. Although every argument which may be derived from this analogy, cannot be too severely scrutinised before we admit the particular conclusion which it may seem to establish, yet we may confidently reckon upon the certainty of its existence, as one of the best guides which we now possess, towards obtaining a more perfect elucidation of the general laws of physiology.

(140.) *Vital Stimulants.* — Life, though at the best of only temporary duration in organised bodies, cannot be maintained in them at all, without the continued application of certain stimulants. All require peculiar kinds of food, according to their respective natures; a sufficiency of air, of moisture, of heat, &c. If entirely deprived of these stimulants, they soon die; and even when they are only partially subjected to their influence, in a less proportion than is requisite for a free exercise of their functions, they languish and become sickly. But, besides the various salutary influences to which all living bodies must be submitted, in order to secure for them a due and healthy performance of their several functions, there are others to which they may be subjected, which are decidedly noxious under all conditions, and which must ultimately prove fatal to them, if they had not the power of escaping from their presence, or at least of modifying their effects. In proportion as a living being possesses a greater power of choice, either in profiting by those circumstances which are favourable, or in avoiding those which are hurtful to its existence, we may consider it to be more elevated in the scale of nature, and further removed from the condition of mere brute matter. Most animals, by the faculty which they possess of locomotion, have a great advantage in this respect over plants; and even those among the very lowest tribes of animals which are permanently fixed to one spot during the whole period of their existence, still possess a certain power of selecting their food, and rejecting what is noxious to them, which vegetables have not. The consequence is, that the continued influence of external agents, is found to be far greater in modifying the characters of plants than of animals. As a sort of compensation however, the vital principle in plants is so much less energetic than in animals, that they are not so readily affected as these latter, under any merely casual or temporary alteration in the external conditions under which they may be placed.

(141.) *Properties of Tissue.* — Before we describe the functions performed by the vegetable tissues, it will be necessary to remark upon a few of the properties which these tissues themselves possess. In the complex phenomena which vegetation furnishes, it is very difficult to separate so much of each result as may be strictly ascribed to the operation of the vital principle, from such as may be due to the action of purely physical causes, the chemical effects of affinity, and the mere mechanical properties of the tissue. The most obvious means which we can employ, for ascertaining the precise properties of the tissue, is to perform experiments upon it in the dead vegetable, and as nearly as possible before any chemical change may have taken place in it. It will not be necessary for us here to notice all the properties which the vegetable tissues possess in common with other substances; but there are two on which we shall make a few remarks, as the phenomena to which they give rise might in some cases be attributed to the operation of the vital force: these are, the elastic and hygroscopic powers of some vegetable tissues.

(142.) *Elasticity of Tissue.* — This property is eminently conspicuous when the tissue is distended with fluid; and, unless its effects be duly appreciated, we might be misled, and inclined to consider certain phenomena as the direct result of an irritability residing in the plant, whilst, in fact, they may be easily accounted for by the action of elasticity alone. Thus, in the flowers of the common nettle (*fig.* 147. *a*), the filaments are at

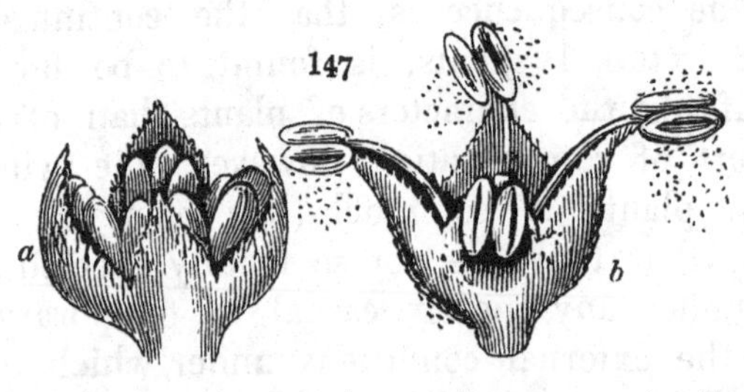

first curved inwards, and the anthers meet in the centre. When the flower is completely expanded, the filaments have become highly elastic; but are still retained in their original curved position by the mutual pressure which they exert upon each other. If this state of equilibrium be disturbed, either by slightly displacing the anthers with the point of a pin, or by the further progress made in vegetation, the stamens are suddenly thrown back by the elasticity of their filaments, the anthers burst and the pollen is scattered by the shock (*b*). This appearance is very like that of some other sudden motions, which, as we shall hereafter show, must be referred to the direct influence of some stimulus upon the vital principle. Many seed-vessels when fully ripe, burst as it were spontaneously, by the increased elasticity of their tissue, and the seeds are often scattered to a considerable distance by this means; but although all the organs of plants when replete with fluid, are generally elastic, a remarkable exception occurs in the pedicels of *Dracocephalum moldavicum*

When these are turned in any particular direction, they retain the position in which they are placed, without any effort to return again to that in which they were previously disposed.

(143.) *Hygroscopicity of Tissue.*—The hygroscopic properties of some tissues are very great, and are the cause of certain motions, which might be mistaken for the direct effects produced by the vital force. If the awn or bristle of the wild oat be moistened, it immediately untwists; the teeth of mosses suddenly collapse when moistened by the breath, and readily expand upon drying again. In estimating the hygroscopic properties of the tissue, we must distinguish between the action of the whole mass, and the property of the membrane which forms the separate vesicles and tubes of which the tissue is composed. It seems easy to account for the hygroscopicity of the mass of the tissue, when we remember that it is penetrated in all directions by intercellular passages, and thus resembles a sponge, which absorbs moisture by the common properties of capillary attraction. This action is found to be much more powerful in proportion as the vegetable tissue is but slightly charged with foreign matter. Some plants, as the mosses, readily imbibe water, however long they may have been dried; and reassume an appearance of freshness nearly equal to that which they possessed in a living state; but, in these cases, the effect is most probably due to the hygroscopic action of the elementary membrane composing the vesicles, and not to the capillarity of the tissue itself. The immediate result of any hygroscopic action upon a portion of the tissue is to enlarge it; and consequently, where two portions are in contact, one of which is more hygroscopic than the other, there exists a tendency to separation. When, however, they do not separate, the portion which is the least hygroscopic, becoming less distended than the other, necessarily produces an incurvation of the mass upon that side on which it is placed.

(144.) *Endosmose.*—Connected with the hygroscopicity of the vegetable membrane, we may here mention a property of all membrane, which has probably a considerable influence in the economy both of animal and vegetable life. When a membrane is viewed under the highest powers of the microscope, it appears to possess a perfectly homogeneous texture, without pores of any kind; and yet water, milk, and other fluids, placed under certain circumstances, are capable of passing through it with considerable facility. The conditions required for producing this effect are these: — Any two fluids which exert a mutual affinity towards each other, being placed on opposite sides of a membrane, their immediate intermixture will commence, each of them passing through the substance of the membrane. If, for instance, a little treacle be enclosed in a piece of bladder, and this immersed in water, a portion of the treacle will soon be found to have exuded, whilst a still larger quantity of water will have penetrated into the bladder; and this action will continue until the fluids have acquired the same density. The remarkable circumstance attending this phenomenon, is the fact of the lighter fluid having penetrated the membrane with greater velocity than the denser fluid. In consequence of this, the bladder becomes distended. By a simple contrivance, styled an endosmometer, we may measure the degree of force or velocity by which the current of water exceeds that of the current of the denser fluid. In *fig.* 148 A is a glass funnel with the mouth downwards, and covered with a piece of bladder.

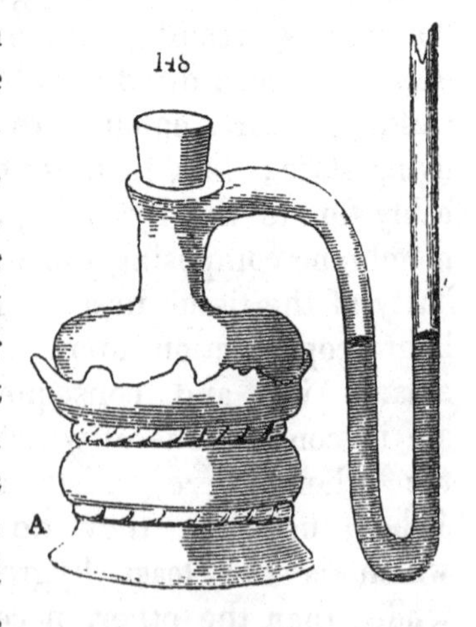

The other end of this funnel is furnished with a tube twice bent, the stems of which are vertical; treacle is placed in the body of the funnel, and the mouth immersed in water; mercury is poured into the open extremity of the tube, and ascends in the other stem until it meets the fluid in the funnel. So soon as the endosmose commences, the rising fluid pushes the mercury before it; and the amount of the force by which this is effected, is ascertained by pouring in more mercury until the further rise of the fluid is checked. The height of the column of mercury affords an estimate of the pressure of the ascending fluid, which is of course due to the force of the endosmose. In this way it may be shown, that a syrup three times the density of water produces an endosmose capable of sustaining a pressure equal to the weight of three atmospheres.

(145.) *Vital Properties.* — After abstracting all that can reasonably be allowed to the physical properties of the tissue, and to the chemical or other effects which operate in modifying every vital phenomenon, whatever still remains unaccounted for in the functions of vegetation, must be ascribed to the direct operation of the vital force itself. What life is, whether it is a simple quality, the effects of which are variously modified according to the nature of the tissue in which it resides, and by means of which it acts, or whether it possesses several distinct properties, which are severally capable of acting only upon and through particular tissues, is quite unknown to us. For the sake of convenience, and provisionally merely, the physiologist considers animal life to be compounded of certain properties, and that its various functions are performed by these properties, acting through the intervention of different kinds of tissue. There are three of these properties attached to animal life, which may be styled respectively its excitability, irritability, and sensibility.

(146.) *Excitability.*—The excitability of animal life, which is also termed the "vis formativa," is manifested through the cellular tissue, by which the function of nutrition is carried on; it is that property by which this tissue takes cognizance of the action of external influences upon it, and by which it resists those mechanical and chemical efforts which otherwise would soon succeed in decomposing its substance. The existence of such a property is equally evident in the vegetable as in the animal kingdom. No one will deny that vegetables live; and we may perhaps believe, that the general law of life by which they resist destruction, is the very same in kind, however different it may be in degree, as that by which animals are also maintained in a state of existence. In animals indeed, the intensity with which this vital property acts is greater than in vegetables; but, as a sort of compensation, we find that vegetables are much more tenacious of life than animals. A plant may be mutilated to a very great extent, and its separate parts will still live, and are frequently capable of becoming distinct individuals; and, although there are certain creatures possessing a compound structure, among the lowest tribes of animals, yet even in them this property does not reside in so eminent a degree as in certain vegetables, every elementary organ of which appears capable of existing in a detached form, and of reproducing an individual, similar to the original of which it formed a trifling and subordinate part. This therefore, the "excitability" of life as it has been termed, is a property which we may consider common to both kingdoms of organised nature.

(147.) *Tenacity of Life.* — A plant may lose nearly half its weight by drying, and yet be restored by care. De Candolle has recorded an instance of a *Sempervivum cæspitosum,* which had been placed in a herbarium for eighteen months, and from which he afterwards detached a living bud and reared a plant. But the tenacity of vegetable life is best exhibited in the property which seeds possess, of retaining their powers of germination after having been exposed to very considerable extremes of heat and cold. Some also, which have partially germinated, may be again dried and kept for months, without losing the power of germinating afresh, although they are sensibly weakened by such treatment. The revival of plants among the cryptogamic tribes, after a very long suspension of the vital functions, is well authenticated.

(148.) *Irritability.*— Besides the excitability of vegetable life, there are certain striking phenomena exhibited by some plants, which seem to indicate the presence of a property analogous to that of animal "irritability." A closer examination, however, of the circumstances under which this "vegetable irritability" manifests itself, rather inclines us to believe with De Candolle, until sufficient proof be brought to show the contrary, that these are only extreme cases of the operation of the property of excitability. The sudden inclination of the stamens in the berberry towards the pistil, when the filaments are touched near the base on the inside, the well-known phenomena exhibited by the sensitive-plant, and several other singular movements of particular organs in some other plants, are the phenomena which have led to the conclusion, that some few vegetables are endowed with an irritability analogous to that which exists in all animals. But on the other hand it has been observed, that in animals this property is confined to the muscular fibre, whilst in vegetables there does not appear to be any particular tissue to which it is peculiarly restricted. In animals, again, the effects of irritability are apparent during the whole course of their life, and are not destroyed by repetition of the experiments by which they are elicited; whereas this property can be traced only under peculiar conditions of vegetable existence, and then only in certain organs of a very few species. Several of these instances, also, are only special modifications of certain actions, which are constantly produced by the operation of more general causes. For instance, the folding of the leaflets of the sensitive-plant, which takes place when we touch them, is the very same sort of effect which we daily witness in a vast number of other plants, where it is elicited by the agency of light, only in a more gradual and imperceptible manner. In these latter cases, the effect is denominated the sleep of plants, and may be more especially witnessed in the leguminose tribes, whose leaves remain folded during a certain portion of the day, and assume an appearance of languor and inaction singularly analogous to the periodical state of repose exhibited in the animal kingdom. In cases therefore, where similar effects are brought about by the action of certain stimuli, in a yet more violent or rapid succession, we may imagine that they are nevertheless the results of the same vital property, which is here exhibited under some peculiar degree of excitement.

(149.) *Examples of Vegetable Irritability.*—As some of the phenomena exhibited by vegetable irritability are very striking, we shall here insert a brief notice of a few of the most interesting examples.

(1.) *Sensitive-Plants.* — There are several species of sensitive-plants, which possess the property of moving their leaves when they are touched, or otherwise stimulated. The most common is an annual (*Mimosa pudica*), with compound digitate leaves, with four pinnules (*fig.* 149.);—each partial petiole being furnished with numerous pairs of leaflets, expanded horizontally as at (*a*). One of the most striking

means of eliciting the phenomenon in question, is by scorching a single leaflet in a candle, or by concentrating the sun's rays upon it with a lens. This leaflet will immediately move, together with the one opposite to it, both bringing their upper surfaces into contact, and at the same time inclining forwards, or towards the extremity of the partial petiole on which they are seated (*b*). Other pairs of leaflets, nearest to the one first stimulated, will then close in succession in a similar manner; and at length the partial petioles themselves fold together, by inclining upwards and forwards. Last of all, the influence is transmitted to the common petiole, which bends downwards with its extremity towards the ground (*c*); in a direction the reverse of those which were taken in the former cases. The effect is next continued to the other leaves nearest to the one first stimulated, and they fold their leaflets and depress their petioles in a similar manner. When the plant is shaken, all the leaflets close simultaneously, and the petioles droop together; but if the agitation be long continued, the plant will at length become accustomed to the shock, and after a lapse of some time, the leaflets expand again. The mechanism by which these movements are produced resides in the thickened or swollen joints, seated at the bottom of each leaflet and petiole; for if the upper part of these swellings are cut away, the leaf remains erect; but if the lower part is removed it continues depressed. Hence it appears that the elevation and depression of the leaf, is owing to the elasticity of the tissue of which the swollen joint is composed; and that the stimulus employed to produce motion, tends to weaken the upper parts of these joints in the case of the leaflets and partial petioles, but the lower part of those belonging to the main petioles — the contrary sides continuing elastic, as before. But how the effect is produced, and what may be the law which regulates its action, is not known. The causes are active from the earliest stages of the plant's existence; the cotyledons themselves exhibiting the property so soon as they have expanded. The transmission of the stimulus from one leaf to another along the stem of the plant, has been shown by Dutrochet to take place through the intervention of the ducts contained in the woody parts. For, if both the pith and the cortical portions are removed, the effects are not stopped; whilst, if the woody parts are abstracted, which contain the ducts, they cease entirely.

(2.) *Desmodium gyrans.* — The Desmodium gyrans is another plant of the same natural order as the sensitive-plants, the motion of whose leaflets is still more striking than in the latter; for here the motion is continued, without the necessity of applying any external stimulus. The leaves are composed of a pair of small leaflets, and a terminal one of larger dimensions (*fig.* 150.). The motion consists of a succession of little jerks, produced at intervals of a few seconds. One of the two lateral leaflets is gradually elevated, whilst the other is depressed; and when both have attained the maximum amount of movement in one direction, they begin to proceed in the opposite. At the same time the terminal leaflet becomes inclined by similar interrupted movements; first on one side, and then on the other.

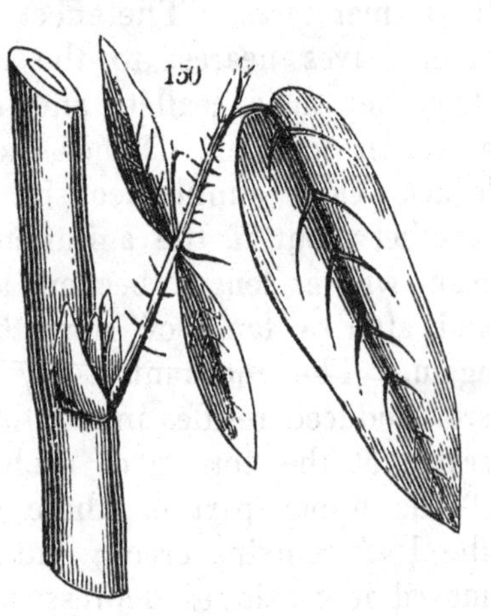

(3.) *Common Berbery.* — The flowers of the common Berbery contain six stamens, which surround a single pistil. When first expanded, the stamens are inclined back upon the petals or away from the pistil. If the filaments are touched near the base on the inside, they immediately start forward towards the pistil, so that the anther is brought close to the stigma. In a little time they recover their original position, and may be again stimulated as before. When the anther is ripe, the violence of the motion causes it to burst, and the pollen is projected on the stigma; and we may unquestionably consider the mechanism by which this effect is produced as designed for effecting this very purpose.

(4.) *Dionæa muscipula.* — The leaves of the Dionæa muscipula, or Venus's Flytrap, consist of a flattened petiole (*fig.* 151. *a*), at the extremity of which are two fleshy lobes (*b*), which lie when expanded in the same plane with the petiole. These lobes are capable of being elevated and brought together into a position perpendicular to the surface of the petiole (*c*). They are furnished with "ciliæ," or bristles, round their margins, which stand nearly at right angles to their upper surface; and there are besides these, three little short bristles placed upon the upper surface of each lobe in a triangular order. When a fly or other insect, crawling over the surface of the lobes, touches either of these latter bristles, the irritability is excited, the lobes suddenly close, and the insect is imprisoned like a rat in a common gin. Some little time after the death of the insect, the lobes unfold and wait for another victim. The only plausible conjecture which has been made, to account for the use and intent of this singular contrivance, supposes this plant to require animal manure for the healthy performance of some function or other; and in corroboration of this opinion, it has been stated that Mr. Knight, after having secured some plants from the possibility of providing themselves with flies, furnished some of them with scraped beef, and left the rest without any such provision. The result of the

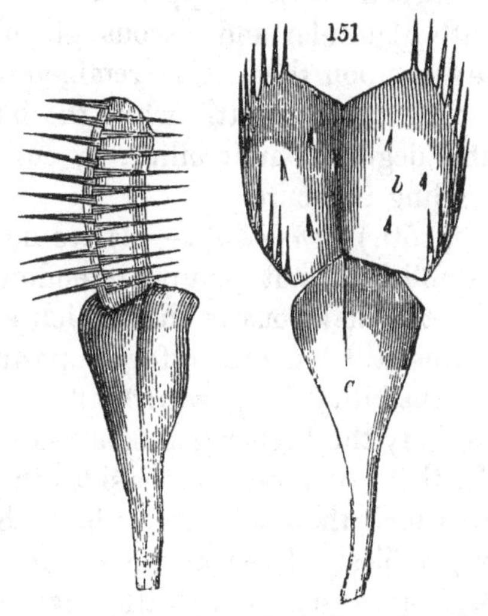

experiment showed the more flourishing condition of the provisioned specimens.

(5.) *Sundews.* — To the above list we may add one more example, taken from a British genus of plants, the Droseræ or Sundews, of which three species are natives of this country. The leaves of these plants are furnished on their upper surface with long hairs, tipped with glandular and viscous globules. When an insect settles upon them it is retained by the viscosity of the gland, and in a little while the hairs exhibit a considerable degree of irritability, by curving inwards, and thus holding it secure.

(150.) *Sensibility.* — If we do not consider it clearly established that plants are endowed with an irritability strictly analogous to that which exists in animals, there seems still less reason for supposing them to possess that "sensibility," by which all animals, but more espepecially the higher tribes, are so eminently characterized. In them this property resides in their nervous system, to which there appears to be nothing analogous among vegetables. Even in the lower tribes of animals, their nervous system is so little developed, that they may be mutilated and otherwise injured, to an extent which would speedily cause their death, if the intensity of the pain which they felt were at all proportionable to what animals of a higher grade experience under similar treatment; and yet they scarcely appear to suffer any inconvenience. If there were no better argument to satisfy us that plants are utterly devoid of sensibility, we have the general consent of mankind, founded on their daily observation, in favour of the non-existence of such a property. The only plausible arguments in support of the probability of plants being endowed with something analogous to a nervous system, rest upon the effects produced on them by different poisons. When corrosive poisons are imbibed into their system, they destroy the tissue much in the same way as in the animal frame; but when narcotic poisons are imbibed, although they kill the plants, they do not appear to have produced any derangement or disorganisation in their tissue. But it has been argued that, as these latter poisons act upon the nervous system of animals, we may suspect something analogous to this system to exist in vegetables also. A long list has been given of substances which act as poisons on plants ; and it has been ascertained that very nearly all such as are deleterious to animal are so likewise to vegetable life, and many others besides, which animals may take with impunity. Some of those which it is necessary to administer in large quantities in order to produce death in animals, are sufficiently powerful to kill plants when given in very small doses — as alcohol, ethers, and oils ; whilst on the other hand, the oxides of lead and zinc, which poison animals when administered in small portions, produce little or no effect on plants, probably because they are incapable of being absorbed by the spongioles. Most vegetable extracts and excretions act as poisons on all plants (even upon those from which they were obtained) when they are imbibed by the roots. Gases diffused in water are harmless. Many salts are highly noxious, but most of the salts of lime produce no effect. Fortunately for the permanence of vegetation on the surface of the earth, the natural poisons are not very generally diffused in places where plants are likely to grow.

(151.) *Periodicity.* — In tracing the various analogies which exist between the phenomena of animal and vegetable life, we find a remarkable example in what may be termed the individual temperament, or idiosyncracy of a living organic being. Besides that general resemblance between the manner in which the same functions are performed by all individuals of the same species, there are certain modifications in the results which are peculiar to particular individuals, and which must be attributed to some peculiarity in their temperament. This is remarkably exhibited in the differences observable among separate individuals of the same species, as regards their periods of leafing or flowering; for although it is evident that the regular return of the seasons stimulates all plants to a periodic execution of these functions, and although the great majority of individuals of the same species and under the same circumstances perform them at nearly the same time, yet it often happens that some individuals are considerably retarded or accelerated in these respects. But further than this, the functions themselves, independently of the action of any external stimuli, appear to have a natural inherent tendency to periodic returns of activity and repose. Thus in the animal kingdom, the return of night and day are met by a desire to sleep and to be awake; and although these desires may be so modified in different individuals that some require less sleep than others, there are certain limits beyond which it is not safe to carry any unnatural attempts to live without it. Now as in these cases we do not attribute the periodic desire to sleep to the regular return of night, but to the character of the function itself; so in the case of the diurnal opening and closing of flowers, the phenomenon must primarily be ascribed to some inherent quality in the plant, assisted indeed by the stated returns of the stimuli to which it is subject.

(152.) *Functions of Vegetation.*—Whether we consider life in the vegetable kingdom as possessing more than one property or not, the various operations which result from its action, upon and through the instrumentality of the several organs of which plants consist, are termed "functions of vegetation." Although there are a multiplicity of operations carried on in different parts of the vegetable structure, they may all be considered subordinate to one or other of the two general functions of nutrition and reproduction. By the former the life of each individual is preserved, and by the latter the continuance of the species is secured.

(153.) *Stimulants to Vegetation.*— Life, in order to act through the instrumentality of the vegetable structure, requires to be stimulated by the influence of external agents. Unless such be present, the vital force remains dormant, even where it is not extinguished. Thus for example, seed will not germinate unless it be placed under peculiar circumstances with regard to moisture, temperature, and the atmosphere ; but when a sufficient supply of these three stimulants is provided, the seed swells, bursts, and the plant is gradually developed. The principal stimulants to vegetation are light, heat, air, and water ; and the conjoint action of at least three of these four is generally requisite to secure a healthy condition to most plants.

(154.) *Light.*— The action of light, as we shall show more distinctly when we are describing some of the functions of vegetation, is of the greatest importance. We shall here notice only one phenomenon, to which we have already alluded (art. 148.), where the presence of this stimulant exerts a decided influence.

(155.) *Sleep of Leaves.*— The phenomenon to which we allude is termed the sleep of plants. This consists in a periodic change in the position of an entire leaf, or of the several leaflets of which a compound leaf is formed. The petioles, or leaf stalks, either bend upwards or downwards, so that the flattened surface or limb of the leaf is elevated or depressed. There are about a dozen different modifications in the manner in which the leaves are inclined to the stalks on which they grow ; some raise their leaflets so that their upper surfaces are brought into contact, and others depress them so that the under surfaces meet together. This phenomenon is best exhibited by various species of the two natural orders, the Leguminosæ (which includes both the pea-flowering plants, as clover, &c., and the acacias and mimosas, &c. which have regular flowers) and the Oxalideæ. These phenomena depend upon a special physiological law, subject in some degree to the stimulating effects of light and heat, which elicit and control them, but which are not themselves the primary causes of these effects. When the sensitive-plants are confined in a dark room, their leaflets periodically fold and open as usual, excepting that the periods are somewhat lengthened; on the other hand, when they are exposed to a continued light, these periods are shortened. When exposed to strong lamplight by night, and excluded from all light by day, their periods of sleep become extremely irregular for a time; but, in the end, the specimens generally close their leaves during the day, and unfold them at night. The alternate opening and closing of flowers is a similar function to that of the sleep of leaves. The time of day in which flowers close is very different for different species, and even differs for that period during which the leaves are asleep on the very same plant. Bertholet mentions an acacia in the garden at Orotava in Teneriffe whose leaflets closed at sunset and unfolded at sunrise, whilst its flowers closed at sunrise and expanded at sunset.

(156.) *Electricity.* — Nothing very decisive is known of the effects which so important an agent as electricity produces on vegetation. It is, indeed, supposed to act as a stimulant, and the supposition is countenanced by the increased vigour with which plants are observed to grow during the prevalence of stormy weather. It seems to be not unlikely, that some trees are more liable to be struck by lightning than others; but they are all so constructed as to present numerous conducting points in the extremities of their branches, well adapted for drawing off the electricity in the clouds.

(157.) *Temperature.* — The influence of temperature on vegetation is a very important consideration, whether we regard the physical or physiological effects which it produces. When the temperature is below the freezing point plants can obtain no nutriment, because the water in which it is conveyed is solidified. But further, it is essential to the healthy condition of every plant that its internal temperature should be supported within certain limits, which differ for different species. The opposite extremes of temperature under which different plants are capable of existing are widely apart. Some flourish within the influence of hot springs, where they are stated to be constantly exposed to a temperature of 62° R., or $171\frac{1}{2}$° F., and even to 80° R., which is equivalent to 212° F.; whilst the oak sustains the rigours of a winter in latitudes where the thermometer falls to –25° R., or $-24\frac{1}{4}$° F., and the birch will resist a cold of –36° R., or –49° F. The latter is well protected against the effects of extreme cold by the manner in which its trunk is defended with several loose coats of epidermis. The chief protection, however, against the sap freezing in the trunks of trees, is the circumstance of its being contained in extremely minute vesicles and capillary vessels; for it has been shown that water will resist a temperature of –7° R. or $16\frac{1}{4}$° F. under similar circumstances; and all viscid fluids are still more difficult to freeze than water. Whenever the sap does freeze, it produces the effect technically termed "shakes" in timber trees, which consists in a tendency in the separate layers of wood to disunite.

(158.) *Internal Temperature.* — In animals, the function of respiration is the means by which caloric is set free, for the purpose of maintaining the temperature of their bodies at a sufficient elevation to protect them against the influence of cold, and perspiration cools them when they are exposed to excessive heat. As vegetables perform two functions of a similar kind, we might perhaps be led to expect that the influence of similar effects would regulate their internal temperature. But, if such be the fact, the results are on too minute a scale to be rendered sensible by our instruments; and in the winter, when these functions nearly cease, we cannot suppose that they operate at all in resisting any atmospheric changes which might be injurious to vegetation. Still it has been observed as a general law, that the temperature of a tree is higher between autumn and spring than the average temperature of the air, and that it is lower between spring and autumn. But there are physical causes which seem to be sufficient to account for these facts without the necessity of ascribing them to the results of any physiological action. The roots penetrate the earth to a depth where the soil is always warmer than the atmosphere in winter and cooler in summer, and the moisture which they imbibe will consequently partake of this influence. Hence it has been observed, that the internal temperature of trees is about the same as the soil at one-half the depth to which their roots penetrate. The maintenance of an internal temperature distinct from the external is assisted by the nature of the wood itself, which is a bad conductor of heat; and also by the property which it possesses of conducting heat better in a longitudinal than in a transverse direction. As an example, we may mention that the milk of the cocoa-nut is kept cool during the hottest part of the day by the thick fibrous coating of the pericarp, which is a very bad conductor of heat.

CHAP. II.

FUNCTION OF NUTRITION — *Periods* 1, 2, 3, 4.

ABSORPTION (160.). — ASCENT OF SAP (163.). — CAUSES OF PROGRESSION (165.). — EXHALATION (169.). — RETENTION OF SAP (172.). — RESPIRATION (173.). — FIXATION OF CARBON (176.). — ORGANIZABLE PRODUCTS — GUM (177.). — ETIOLATION (179.). — COLOURS AND CHROMATOMETER (182.). — RESULTS OF RESPIRATION (189.).

(159.) *Function of Nutrition.* — THE first of the two general functions (art. 152.), that of nutrition, may be conveniently subdivided into about seven distinct processes or subordinate functions, which are all carried on simultaneously in different parts of the vegetable structure, more especially during those seasons of the year in which the powers of vegetation are the most active. Sometimes, only one of them is in activity, whilst the rest are either partially or entirely suspended. But as the whole of the materials which serve to nourish the plant must have been subjected to these several processes in succession, we may consider the function of nutrition to be carried on during as many successive periods, before it is completed. We will briefly mention what these successive processes are, before we enter upon the details necessary for the more accurate description of each of them. In the first place, plants absorb their nutriment by the roots; this nutriment is then conveyed through the stem into the leaves; there it is subjected to a process by which a large proportion of water is discharged; the rest is submitted to the action of the atmosphere, and carbonic acid is first generated, and then decomposed by the action of light:

carbon is now fixed under the form of a nutritive material, which is conveyed back into the system; and this material is further elaborated for the development of all parts of the structure, and for the preparation of certain secreted matters, which are either retained within or ejected from the plant. These several processes may be designated: 1. Absorption; 2. Progression of sap; 3. Exhalation; 4. Respiration; 5. Retrogression of proper juice; 6. Secretion; 7. Assimilation.

FIRST PERIOD OF NUTRITION.

(160.) *Absorption.* — That plants absorb moisture from the soil in which they grow admits of easy proof. The extremities of the fibres in which their roots terminate, are not covered with an epidermis like the rest of the surface, and consequently the cellular texture is there exposed, and constitutes the "spongiole," or true absorbing organ. As plants do not possess the power of locomotion, it is essential that their food should be so universally distributed that they may run no risk of perishing from want of a constant supply. It is further requisite that their food should be offered them in a fluid form; for it is an established principle in vegetable physiology, that the spongioles are incapable of absorbing any matter in a solid state. Whatever therefore, is to be received into the system for the purpose of nutrition must be held in a state of solution in water. The three most important ingredients to be found among the products of vegetation, are oxygen, hydrogen, and carbon (see art. 14.); the two former are the elements of water, and the third is an element of carbonic acid, a gas which is every where present in the atmosphere, and which may be detected in almost all springs and other waters on the surface of the earth. Water, again, in a state of suspension in the air, is also present every where. Plants, therefore, receive a constant supply of these three elements wherever they are placed on the surface of the earth, in situations adapted to their growth. Besides the three elementary substances, oxygen, hydrogen, and carbon, essential to the composition of all organized matter, whether animal or vegetable, there are other elements to be met with in slight proportion in some vegetables. Azote is an element more especially essential to the formation of animal substances; but it seems probable, that it is also a fundamental ingredient in certain vegetable compounds, in which it exists in considerable abundance. As this gas also forms a component part of the atmosphere, plants may as readily be furnished with it, as with either of the other three ingredients universally essential to their nature. Whether the other elements occasionally found in plants ever constitute an essential part of their structure, is uncertain. Several of them exist under combinations, such as common salt for example, which appear to be useful to some plants; possibly as a stimulus necessary for the preservation of their health, since they languish and die when wholly removed from their influence. In all cases, however, whatever be the nature of the various saline, earthy, metallic, and other compounds found in small quantities in the ashes of plants, they must have been introduced in a state of solution through the spongioles.

(161.) *Cause of Absorption.* — This absorption by the spongioles continues during the lifetime of the plant, and it becomes a question for the physiologist to determine, upon what cause the action depends; whether it may be ascribed, for instance, to the known hygroscopic powers of the cellular tissue, or whether it be wholly or partly due to a vital action. This question can scarcely be considered as satisfactorily settled. If we suppose the plant capable of removing the imbibed fluid as fast as it is absorbed by the spongioles, then we may imagine the possibility of a supply being kept up by the mere hygroscopic property of the tissue, much in the same way as the capillary action of the wick in a candle maintains a constant supply of wax to the flame by which it is consumed. This view is further supported by the fact, that the facility with which different liquids are absorbed, appears to depend entirely upon their degrees of fluidity; and thus even the most noxious materials will be more readily imbibed than such as are nutritious, provided they are presented to the spongioles in the more fluid state. Now if their absorption were the result of a vital action, we might have expected that a greater degree of energy would have been exerted in favour of the more nutritious matter, and that the noxious ingredient would have been absorbed with difficulty.

(162.) *Stimulants to Absorption.*—Whatever be the immediate cause of absorption, it does not depend upon the action of light; for plants absorb by night as well as by day, and the absorbing organs are most frequently placed under ground, and in the dark. In an indirect manner, however, light does certainly exert a considerable effect upon the quantity of fluid absorbed; because it is the stimulant by which a large portion is continually removed by the function of exhalation; and we consequently find that when plants are placed in the dark, although the absorption continues it is considerably checked, so that the water imbibed accumulates until they become dropsical, and their leaves fall off upon the slightest touch. An increase of temperature augments the quantity of water absorbed; but this again may depend upon some local stimulus upon another function. Thus if a branch from a plant growing in the open air be introduced within a stove during the winter, it will immediately begin to push its leaves, and become the remote cause of accelerating the absorption of the sap, which had been going on very languidly.

SECOND PERIOD OF NUTRITION.

(163.) *Ascent of the Sap.*—The fluid introduced by the absorption of the spongioles bears the general name of sap or "lymph." Essentially, this sap is nearly pure water; but in order that it may become effective in nourishing the plant, it must contain carbonic acid, or at least some carbonaceous material capable of being converted into carbonic acid by a subsequent process, which we shall presently describe. In Dicotyledonous woody stems, it has been clearly ascertained that the course of the sap is up the woody fibre, and especially through the alburnum, but that it does not ascend in any appreciable quantity through the pith or bark. It is then carried onward through the branches and into the leaves. In the internal parts of old trunks, the sap accumulates in large quantities about the spring of the year, and is there retained under a certain degree of compression; for if the tree be felled at this season, it flows most readily from those central parts which have ceased to possess any vitality, and sometimes it even issues in a jet during a few seconds, when the trunk is first severed. Whether or not any distinct modification takes place whilst the sap is moving onward, analogous to the effects of digestion in animals, has not been clearly ascertained. It is certain, indeed, that if a tree is tapped at different heights, when the sap is rising with the greatest energy, the liquid obtained from the lower parts of the stem is purer than that which is derived from the upper parts. But this may be ascribed to the complete admixture which takes place between the juices previously elaborated and the ascending sap, which thus becomes thickened by them as it moves onward.

(164.) *Channels for the Sap.*—Some authors suppose the sap to be propelled through the vascular system, whilst others consider it to rise through the intercellular passages, and others again imagine that it passes from cell to cell, through the elementary membrane of which they are formed. The great difficulty in determining the precise channel through which the progression of the sap takes place, must be ascribed to the perfect transparency of the vegetable membrane, and the extreme minuteness of these organs themselves. By placing a branch in coloured fluids, such as a decoction of Brazil-wood or cochineal, they are absorbed and the course of the sap through its whole passage into the leaf may be readily traced; but on examining microscopically the stains which have been left, it is scarcely possible to feel satisfied whether they are on the outer or inner surface of the vessels and cells which they have discoloured. The mutilated state of the stem, when subjected to experiments of this description, has also introduced errors into the results, and the coloured liquids have been observed to rise up certain vessels which under ordinary circumstances appear destined to convey air. Since there are many plants which possess no vascular structure, in them at least we must allow the cellular tissue to be the true channel through which the sap is conveyed. But whatever may be the manner in which the effect is produced in the more succulent parts of plants, it seems to be unquestionable that a more direct mode of progression than that of a gradual transmission from cell to cell, must exist in the older parts of woody stems. If for instance we take a long branch of the vine and bend it in the middle, the sap immediately exudes at the extremities, but chiefly on those sides which are towards the concave surface produced by the flexure; which not only indicates a continuity, but also a rectilinear course in the channels through which the sap is conveyed. It is further evident that a general intercommunication must subsist between these several channels; for the stem may be notched to the very centre, at different altitudes and on different sides, so as completely to intercept every rectilinear communication between the lower and upper parts, and the sap will still find its way into the leaves. The probability therefore seems to be, that the crude sap really rises, at least in woody stems, through the intercellular passages, where it bathes the surface of the cells and vessels, all of which are so many distinct organs destined to act upon it—and more especially when it has afterwards become intermixed with the proper juices of the plant. If this view of the subject should prove correct, then the intercellular passages must be considered analogous to the stomachs of animals, mere recipients of a crude material, which is afterwards modified and rendered available for the purposes of nutrition.

(165.) *Cause of Progression.* — The progression of the sap appears to be influenced by several causes. De Candolle supposes it to be carried forward through the intercellular passages by successive contractions and dilatations of the cells. But there appears to be no warrant for the supposition; on the contrary, it seems impossible that such an effect could be produced in cells which are replete with an incompressible fluid. If contraction were to take place, an expulsion of the contained fluid must ensue, and every dilatation of the cells would require that the ambient fluid should enter them. Whether therefore the sap rises or not through the intercellular passages, the hypothesis which he has framed to explain its progression appears to be inadmissible.

(166.) *Propulsion of the Sap.* — The first and most important cause of the rise of the sap, resides in the spongioles. The water imbibed by them, is also by them propelled forward with considerable force, and the effects are strikingly analogous to those exhibited by the endosmometer (art. 144.). Hales cut off the stem of a vine in the spring, when the sap rises with the greatest velocity, and luted a tube to the top of the stump, bent in the manner we have described in the construction of the endosmometer. As the sap rose into the tube, mercury was introduced at the open end; and a measure of the force of the rising sap was thus obtained, and found to equal the pressure of an atmosphere and a half. If a piece of bladder be tied over the surface of a vine-stump, when the sap is rapidly rising, it soon becomes tightly distended, and will ultimately burst. These effects manifestly bespeak an action very different from the ordinary results of capillarity, and indicate the presence of a powerful force, a "vis à tergo," residing in the lowest extremities of the roots by which the propulsion of the sap is regulated. Although these results so closely resemble those of endosmose, there still exists a difficulty in connecting the two phenomena; for whilst we may admit the possibility of an interchange between the contents of the vesicles composing the spongioles, and the water in the soil which surrounds them, by the ordinary operation of endosmose, it is difficult to explain how the sap may be propelled forward so violently as it appears to be, in the open channels through the centre of the stem, which contain crude sap of nearly the same specific gravity as water itself. It would be further necessary to account for the manner in which a continued supply of fresh materials is obtained for carrying on the endosmose, which must otherwise soon cease when the fluid within has become much diluted. We shall find, however, that a constant supply of fresh material is actually provided by the direct action of the vital force, during a subsequent period in the function of nutrition; and hence it is not impossible, though it has not been proved, that both the propulsion as well as the absorption of the sap may principally if not entirely be owing to the operation of mechanical causes; dependent however for their lengthened continuance upon the existence of the vital energy by which those conditions are perpetually renewed, and without which the endosmose would of necessity soon cease. Although therefore it is quite evident that the immediate effects of the vital force must be somewhere present, and co-operative with the two phenomena we have described, these themselves may be only the secondary results, and not the direct effects of its action.

(167.) *Adfluxion.* — Another cause which promotes the rise of the sap, is the continued discharge of moisture which takes place from the surface of the leaves and other parts, by a process to be described presently (art. 168.). This effect produces a constant absorption from below; and thus a branch placed in water gradually imbibes a large quantity at its cut extremity. This "adfluxion" of the sap, as it has been termed, is clearly the result of a different cause from that of its propulsion, explained in the last article.

THIRD PERIOD OF NUTRITION.

(168.) *Exhalation.* — A large portion of the water imbibed by the spongioles is afterwards discharged at the surface of the leaves, in a manner analogous to the insensible perspiration of animals. This discharge may be attributed to the operation of two distinct causes. A very small portion is carried off by the ordinary effects of evaporation, but a far greater quantity by a process which has been named "exhalation," and which is ascribed to the immediate action of the vital force. That a certain portion of the discharge must be due to the evaporation of the contained fluid through the membranous coats of the vesicles, is proved by the gradual desiccation of the succulent parts of dead plants, and by the effects observed in the preservation of pulpy fruits. But still, the effects of evaporation alone are scarcely perceptible, when compared with the rapid manner in which the fluid is discharged from the surface of the leaf. It has been ascertained that a common sunflower of three feet in height, will exhale about twenty ounces of water every day; and a common-sized cabbage discharges moisture at the same rate: so that the surfaces of these plants exhale at a rate which is seventeen times greater than that at which the insensible perspiration is given off from the surface of the human body.

(169.) *Exhaling Organs.*—By comparing the effects produced by the leaves of different species, it has been found that those exhale the most which possess the greatest number of stomata; whilst those surfaces which possess none, produce very little or no effect beyond the ordinary loss sustained by evaporation. It is quite as evident therefore that the stomata are the true exhaling organs of plants, as that the spongioles are their real absorbing organs. As the under surfaces of leaves are in general more plentifully supplied with stomata than their upper surfaces, the exhalation is there the most abundant. Plants which live under water have no stomata; but as they have no true epidermis either, they rapidly fade when exposed to the air, from the more decided effects of evaporation alone.

(170.) *Stimulants to Exhalation.* — The manner in which the stomata act is unknown; and consequently we are compelled to ascribe the function which they perform to the immediate operation of the vital force. The stimulus by which their activity is sustained, is mainly if not entirely due to the influence of light; for the exhalation ceases when the plant is carried into a darkened chamber, and is restored upon its return to the light. Even lamplight is, to a certain extent, sufficient for maintaining this action. The effects of exhalation are remarkably apparent about sunrise, when the temperature is low, and the moisture exhaled is not readily carried off; it then accumulates, and is deposited in innumerable drops upon the surface and edges of the leaves, and is generally mistaken for the effects of dew: but as it collects equally on plants which are under shelter as on those which are openly exposed, this cannot be the true cause. It is by no means clear that an elevation of temperature has any effect in modifying this function; but since it undoubtedly increases the quantity of the evaporation, it becomes difficult to decide whether any portion of the result is due to an increased exhalation also. The manner in which the direct rays of the sun act in stimulating this function, is well known to those who are aware how necessary it is in order to preserve the beauty and freshness of a nosegay, to keep it constantly in the shade. There are certain succulent plants which possess so few stomata that they may be preserved out of the ground for many days and even months, without perishing from want of moisture; and it will frequently happen that Sedums, and other plants of this character, will even push considerable shoots whilst placed under pressure, when preparing for the herbarium: such specimens should first be killed by immersion for a few seconds in scalding water. As juicy plants require most light to secure for them a regular discharge of moisture, we may mention as a piece of practical information, the propriety of exposing as many leaves as possible in the melon frame to the action of the sun's rays, at the same time providing against the accumulation of moisture in the confined situation in which such plants are placed.

The operation of transplanting should be carried on either in the spring or autumn, when plants are destitute of leaves; otherwise the exhalation is too strong at a time when the absorption has been checked, owing to injury sustained at the root. Provided the plants are well watered, the latter inconvenience may to a certain extent be obviated. The water exhaled is so nearly pure, that scarcely any trace of foreign matter is discoverable in it, certainly not more than would be found in distilled water prepared with the greatest care. Even that which is exhaled by aromatic plants is scarcely tainted by any odour. The stomata are in fact the most perfect and delicate stills to be met with in the laboratory of nature.

(171.) *Retention of Sap.* — About two thirds of the fluid imbibed by the spongioles is thus exhaled by the stomata, and consequently about one third must be still retained in the plant. As this portion now includes all the saline, earthy, carbonaceous, and other materials, which happened to be dissolved in the sap when it was first absorbed, the obvious effect produced by the exhalation is to condense these matters, so that the sap becomes a comparatively denser fluid than it was before. As many of the materials thus introduced are not adapted to the purposes of nutrition, they are deposited in those parts where the exhalation has been going on; but the various carbonaceous materials, furnished chiefly by decomposing animal and vegetable substances, are brought into a situation favourable for receiving a peculiar modification, which we shall describe in the fifth period of nutrition. Of the three elements more especially essential to the composition of all vegetable matter, we find that two of them, the oxygen and hydrogen, may be furnished by the water retained after the process of exhalation has been completed.

FOURTH PERIOD OF NUTRITION.

(172.) *Respiration.* — The first actual change produced in the sap is effected by a process analogous to animal respiration. The air is inhaled by the leaf and the fresh surfaces of other parts of the plant, and its oxygen then unites with the carbonaceous matters contained in the sap, and the result is the formation of carbonic-acid. The greater part of this gas is then held in solution by the sap; and the whole or very nearly all the azote which was separated from the oxygen, is exhaled. Besides the carbonic acid thus formed by the plant itself, the trifling proportion every where found in the atmosphere is also inhaled; and a still larger quantity is introduced in the water absorbed by the spongioles. Hence it appears that a threefold provision is made for maintaining a supply of this necessary ingredient. So long as plants remain in the dark, no fresh change takes place in this condition of things; the carbonic acid is retained, but is not fixed in the form of an organic compound. This further result requires the additional stimulus of light, and then the decomposition of the carbonic acid is effected, the carbon becomes fixed under the form of an organisable compound, which we shall presently describe (art. 176.), and all or nearly all the oxygen with which it was united, is exhaled into the atmosphere. So long then as plants continue to vegetate in the dark they tend to vitiate the atmosphere by abstracting its oxygen, and also by the emission of some portion of the carbonic acid which they generate; but when they are exposed to the light, they not only restore the oxygen which they had previously abstracted from the atmosphere, but also give out another portion of this gas, which they set free by the decomposition of the carbonic acid contained in the air, as well as that which was in the water imbibed by the spongioles. In animal respiration, the carbonic acid is immediately expelled from the lungs as soon as it is formed, and the function is then considered complete; and perhaps it would be more logical to divide the function of vegetable respiration into two processes, one of which should comprise the formation, and the other the decomposition, of carbonic acid.

(173.) *Formation of Carbonic Acid.*—The formation of carbonic acid takes place in the leaf, beneath the epidermis; but whether the air penetrates through the stomata or not, is still uncertain. That it cannot universally be introduced through these organs is apparent, since many leaves have no stomata; and in these cases at least, the action takes place through the intervention of the delicate membrane of which the vesicles of the cellular tissue are composed. If a section perpendicular to both surfaces of a leaf be examined under the highest powers of the microscope (*fig.* 152.), the interior will be observed to be chiefly made up of cellular matter, or "parenchyma," whose vesicles are loosely aggregated, so that large intercellular passages exist in communication with each other, through its whole substance. That these passages are filled with air is readily shown by placing a leaf under water, and beneath the receiver of the air-pump. Upon exhausting the receiver, the air contained in the leaf will be seen to escape through the

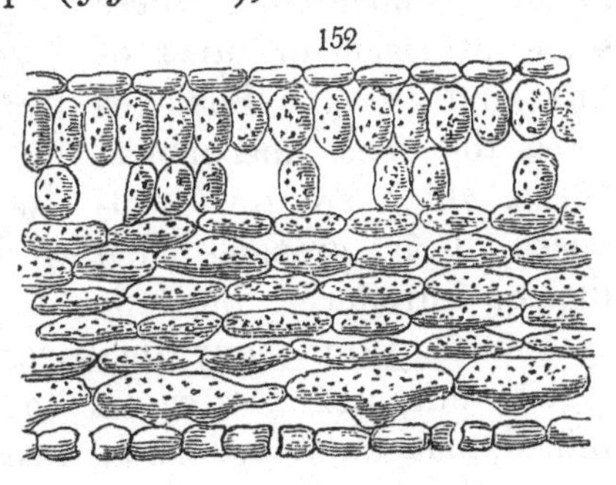
152

petiole; and upon removing the receiver, the water will then find its way into the leaf, and occupy the interstices which were originally filled with air. This effect is rendered particularly striking in those leaves whose under surfaces are of a paler colour than their upper, in consequence of the larger dimensions of the intercellular passages in those parts. When the water is introduced and occupies the whole of these passages, the two surfaces become equally coloured.

(174.) *Air Cells.* — Besides the air in the leaves, some also is found in the stems and other parts of plants, where its precise use has not been fully ascertained. In many aquatics, indeed, it is contained in large cavities, termed "lacunæ," as we have stated (art. 21.). The obvious use of such reservoirs as these, is to float the leaves and other parts in which they exist. The *Pontideria crassipes* has its petioles (*fig.* 153 *a.*) remarkably distended with air. The roots of the Utriculariæ are furnished with a multitude of little bladders (*fig.* 32.) by which they are floated to the surface during the season of flowering; and a number of other instances might be mentioned where some provision or other of this kind exists. But, besides the mere mechanical effects which are thus produced, it is probable that the air introduced into the system may in many instances serve some physiological purpose. It seems to be sufficiently ascertained, that some portions at least of the vascular system are destined to convey air from one part of the plant to another. The spiral vessels and some ducts are often found filled with it; and in these positions, according to some experimenters, it contains rather more oxygen

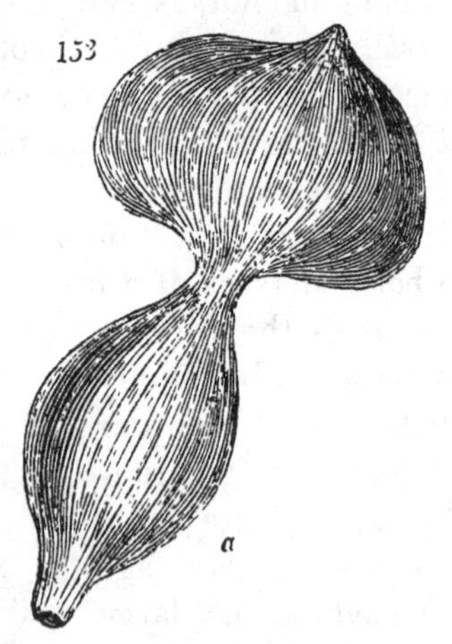

than the atmosphere. At present so little has been ascertained of the conditions under which this air has been introduced into the vessels, or of the peculiar office which it is destined to perform, that we can do no more than just mention the fact, and state the opinion of some botanists, who have considered it probable that in these situations also it is subservient to the process of respiration, and who conclude that it is not impossible there may exist a strong analogy between the manner in which this function is performed by plants and by some of the inferior tribes of animals. Insects for example breathe by introducing air through several spiracles ranged along each side of their abdomen, and which open into certain ducts or pipes, singularly resembling in their general appearance the tracheæ or spiral vessels of plants.

(175.) *Fixation of Carbon.* — When all those parts of plants which are capable of assuming a green tint, but more especially the leaves, receive the stimulus of light, they immediately decompose the carbonic acid contained in the sap. The result of this action is the retention of the carbon, and the expiration of the greater part of the oxygen into the surrounding atmosphere. The most obvious effect produced by this fixation of carbon is the appearance of that green colour which we find in nearly all leaves, and in some other organs. In the few cases which militate against this rule, we may reasonably imagine the existence of some other cause in operation which speedily modifies the initial result. Thus for instance, the peculiar tinge assumed by the leaves of the red-beech, may possibly be owing to the presence of an acid secreted simultaneously with the fixation of the carbon, which converts the green to red. The fixation of the carbon by plants appears to be the first step in that elaborate process by which brute matter is converted into an organisable compound; that is to say, into a material capable of being afterwards assimilated into the substance of an organised body.

Many effects, popularly ascribed to the action of air, are in fact due to the agency of light. Thus trees which grow in elevated or in isolated situations, are more vigorous than others of the same species which grow in forests or in shady places; and those on the skirts of a wood are finer than those in the interior. When fields are arranged into alternate strips of fallow and crop, the produce is much greater from a given portion of land than where the whole field is regularly sown, and this effect must be attributed to the increased influence of light in such cases. The loss of light in stoves and green-houses, by diminishing the effects of exhalation, renders plants more liable to be frozen than others of the same description which are growing in the open air.

(176.) *Organisable Products.*—When we proceed to inquire in what form the carbon appears after it has become fixed, the subject assumes a degree of uncertainty, which it seems almost hopeless to get rid of in the present state of our knowledge. Since this fixation is effected by the leaf and other green parts of the plant, it is consequently in them that we may expect to find the organisable product, whatever it be, which is the primary and immediate result of this action. Now unluckily for our inquiry, there are so many different compounds contained in solution among the sap and various juices of plants, — such as gums, sugars, resins, oils, acids, alkaloids, &c., all of which are composed of different modifications of the same three elements, carbon, oxygen, and hydrogen, — that it becomes a task of the greatest delicacy to determine which of them ought to be considered as the immediate result of the process of fixation. If we may presume that this result is the same in all plants, or so nearly the same that we may designate it (like the blood of animals) by some name which embraces all the subordinate modifications, we must expect to find it among those products which are the most generally dispersed in vegetables, and which are also known to be eminently beneficial to them. These requisites will at once exclude a large class of compounds, to be met with only in certain families of plants, as well as several others which are known to exercise noxious effects upon vegetation. And thus we find, upon careful inquiry, that our choice is restricted to about four substances, all of which possess nearly the same chemical characters, and which are the most universally present among the juices of plants. These are gum, sugar, fecula, and lignine. The first of these appears by far the most universally diffused, and has been obtained from nearly every plant in which it has been sought for; and moreover as it possesses decidedly nutritious qualities, it may be considered with every probability in its favour, as the first or proximate organisable compound formed by the action of vegetable life, acting under the stimulus of light. The other three substances, which so nearly resemble gum in chemical composition, appear to be slight modifications of it, which have resulted from some further elaborations perfected by the vesicles in different parts of the vegetable structure, and we shall defer their description to our account of the sixth period of nutrition.

(177.) *Gum* exudes naturally from certain trees, and especially from some acacias, which furnish the common gum-arabic of commerce. It is purer when obtained in this way than when it has been separated by some chemical process from the sap. Its specific gravity varies from 1·316 to 1·482. It is extremely soluble in water, but is insoluble in alcohol, ether, and and oil. It possesses slight modifications in its qualities, according as it is extracted from different plants; and the following analysis will show its composition, as it has been stated by three eminent chemists: —

	Thénard.	Berzelius.	Prout.
Carbon - -	42·23	41·906	41·4
Oxygen -	50·84	51·306	52·1
Hydrogen -	6·93	6·288	6·5

For the present then, we may consider this substance as most probably the material which is primarily prepared for the nourishment of all parts of the vegetable structure, and which is afterwards further modified by the different vesicles and glands distributed through the system, according as the nature of different parts may require.

(178.) *Etiolation.*— When any part of a plant capable of decomposing carbonic acid is entirely excluded from the light, it remains white. This "etiolation," as botanists term the phenomenon, consists in a combination of an excess of water with the vegetable matter previously prepared; so that the quantity of carbon already fixed becomes as it were diluted, and diffused over a wider space. If the etiolated parts are exposed to the light, the green colour makes its appearance in less than eight-and-forty hours, and the plant gradually assumes a natural and healthy character. The parts which have once become green are incapable of being completely etiolated afterwards. Among the various vegetable matters used by man as food, those which are the least sapid are among the most alimentary; whilst the more highly flavoured are generally more or less deleterious, and some of them extremely poisonous. In order to obtain a food which shall be both wholesome and grateful, the horticulturist contrives by varying his mode of culture to moderate the proportion in which the deleterious ingredients are naturally secreted, and thus renders them harmless. The most common mode of producing this effect is by removing the stimulus of light from such parts as are intended to be eaten; this both diminishes the activity of the organs employed in secreting the deleterious matters, and at the same time causes them to absorb a superabundant supply of moisture. In this way the blanched stems of celery, which in its natural state is a poisonous plant, become a grateful food. The leaves of the endive, and many others which would be far too bitter or tough in their natural state to be eaten, are rendered useful and agreeable additions to our salads.

(179.) *Action of Sun's Rays.*—Although the decomposition of carbonic acid by the green parts of plants, is perpetually carried on under the stimulus of diffused light, and its effects may even be rendered apparent by the action of lamp-light, which gives a slight tinge of green to plants when grown in a cellar, yet in these cases the process is carried on too slowly to allow of our collecting the oxygen which is set free. But when plants are placed in the direct rays of the sun, the action is so much more rapid, that the oxygen may then be collected in sufficient quantity to produce a striking result. If a plant be immersed in pump water, under an inverted glass jar placed in the direct light of the sun, in a short time the surface of its leaves becomes covered with minute bubbles, which presently collect at the top of the jar, and are found to be nearly pure oxygen. When boiled or distilled water is used from which all the carbonic acid has been expelled, no such effect takes place. But if another jar filled with carbonic acid be also inverted over the same pan in which the jar containing the plant is placed, and the surface of the water in the pan protected by a coat of oil, to prevent the escape of the gas as it is gradually imbibed by the water, it will then be decomposed as before, and the oxygen will collect in the upper part of the jar which contains the plant, whilst an equal bulk of carbonic acid will disappear from the other jar. It does not appear that the epidermis is essential to the success of this experiment, and the decomposition of the carbonic acid is equally effected by leaves which have been deprived of it.

(180.) *Action of Oxygen.*—A certain portion of free oxygen is necessary for the formation of the carbonic acid generated by the process of respiration; but when this carbonic acid is decomposed and the carbon fixed, the same oxygen which is set free, will serve again for a fresh formation of carbonic acid so long as there remains any carbonaceous materials in the sap. This may assist us in explaining an interesting fact described in the "Gardener's Magazine," vol. x. p. 208. It is there stated that many plants, especially ferns, have been readily grown in the smoky atmosphere of London, by placing them in boxes furnished with glass coverings hermetically sealed. In this state they have lived and increased in size during several years, without any immediate communication with the atmosphere. The same mode of treatment has been successfully practised in transporting plants during a long voyage, the influence of the sea breeze charged with saline particles forming the greatest obstacle to their safe conveyance. When performing experiments to ascertain the decomposition of carbonic acid by the process of respiration, great precaution is necessary to ensure accurate results. The plants being placed under conditions which are not strictly natural, are soon apt to become sickly and exhibit a tendency to decompose. When this is the case the formation of hydrogen, water, and other substances takes place, and vitiates the results. Those who are anxious to pursue these researches in further detail may peruse the admirable treatises of De Saussure and Ellis, where they will find a multitude of experiments recorded and a patience of investigation exhibited, which has been rarely surpassed by the labours of other philosophers.

(181.) *Vegetable Colours.*—Not only the green colour of those parts which decompose carbonic acid, but all the various colours of plants, depend upon the presence of minute grains of matter contained in the vesicles of the cellular tissue. The grains which give the green tinge to the leaf are termed "chromule," and it is probable that all the others are only modifications of the same substance. From observations made upon the leaf at different seasons of the year, it appears that towards autumn this organ ceases to give out oxygen by day though it continues to imbibe it by night; and hence it seems highly probable that the chromule passes into different states of oxidation, each of which possesses some peculiar tint, as in the case of the various oxides of iron. Although carbon is the principle ingredient in the composition of chromule, it is not likely as some have supposed to be this substance in a perfectly pure state. Although different colours in plants appear to depend upon that action of light which effects the decomposition of carbonic acid, yet we find that many sea-weeds are intensely coloured when they grow at a depth where the illuminating power of the sun's rays is some hundreds of times less than it is at the surface of the earth. Humboldt mentions having obtained the *Fucus vitifolius* from a depth of 190 feet, where the light which it received was two hundred and three times less than that of a common candle placed at the distance of one foot from the object illuminated. All white flowers are only different tints extremely diluted — a fact of which the celebrated flower painter Redouté availed himself. By placing the flower on a white sheet of paper he could observe the exact tint, however delicate, which ought to form the ground of his drawing. All blacks on the other hand are only intense shades of some of the darker colours, or of grey.

(182.) *Colours of Flowers.* — Colour is (generally speaking) of very little importance as respects the determination of species among flowering plants; but it often furnishes characters of considerable value for the discrimination of many among the cryptogamic tribes. In some other branches of natural history it is of much greater consequence; and we shall here explain a method by which an accurate and comprehensive nomenclature may be established for defining colours, so far as may be required in the description of objects of natural history. The scheme is little more than a modification of a plan suggested by M. Mirbel; and consists in referring all natural colours to certain absolute tints and shades*, determined according to fixed rules.

(183.) *Composition of Colours.*—All colours may be

* By "shade" we here mean the depth or intensity of a tint.

referred to different degrees of mixture between three colours, which are considered as "primary." These we may *assume* to be red, blue, and yellow. A mixture of red and blue makes purple; of red and yellow makes orange; of blue and yellow makes green; and innumerable binary compounds may be formed by uniting the primaries two and two in different proportions. Innumerable shades also of each of these may be obtained, between the deepest that can be formed and the faintest, by diluting each colour to a greater or less extent. In order that we may consider every colour to be formed on some regular principle, we divide a circle into three equal parts (*fig.* 154. *innermost*), and place the Blue (B), Red (R), and Yellow (Y), in each of them respectively. Around this circle a second is described, and divided into six equal compartments containing respectively the three primaries, and also those three binaries which are *exactly intermediate* between them; viz. the Orange (R + Y), the Purple (B + R), and the Green (B + Y); assuming these also of the same shade as before. Another circle containing twelve equal compartments is described round the last, and in these are placed the last six colours, together with six new ones formed by uniting each contiguous pair in the same way as before. Another circle would contain twenty-four colours and so on; each fresh addition being always formed from the combination of two contiguous colours in a former circle, and between which it is to be exactly intermediate; and the whole is then reduced to a uniform shade. By proceeding in this way it is evident that we may form every conceivable binary compound, or

154

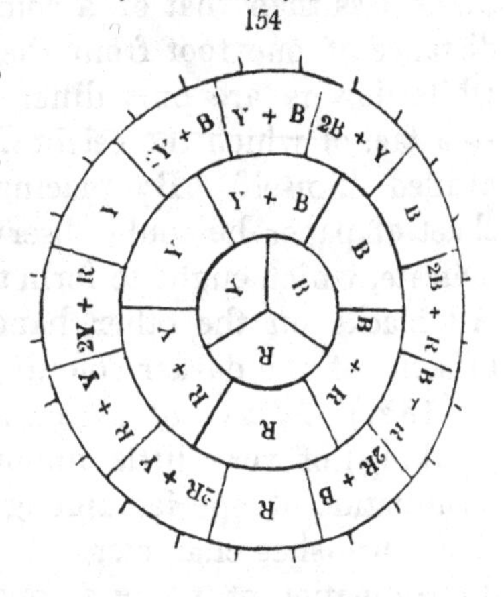

"pure colour." But as the colours in contiguous compartments will differ less and less from each other as we extend our circles, it will not be necessary that we should proceed further than we are able *readily* to appreciate their difference. Now it is considered that the third circle of twelve colours will satisfy the required purpose, and these we name the fundamental or "basial" colours of our scheme. Their composition is expressed in our diagram (*fig.* 154.), and the usual names employed to designate them would be —

B.	Blue.
2 B + R.	Bluish Purple, or Purplish Blue.
B + R.	Purple.
2 R + B.	Reddish Purple, or Purplish Red.
R.	Red.
2 R + Y.	Reddish Orange, or Orange Red.
R + Y.	Orange.
2 Y + R.	Yellowish Orange, or Orange Yellow.
Y.	Yellow.
2 Y + B.	Greenish Yellow, or Yellowish Green.
Y + B.	Green.
2 B + Y.	Bluish Green.

(184.) *Pure Colours.* — It may be here observed that if the three colours purple, orange, and green, or any other three taken at equal intervals round a circle constructed on the above principle, had been *assumed* as our three primaries, and these had been combined two and two, we should have obtained all the pure colours as before, and among them the three former primaries (blue, red, and yellow) under the character of binary compounds. This will be apparent when we recollect that the union of three primaries in equal proportions forms white light with the colours of the spectrum, and a grey or neutral tint (N), when material colours are employed.

Now, Green + Orange = (B + Y) + (R + Y) = (B + R + Y) + Y = N + Y.
Orange + Purple = (R + Y) + (B + R) = (B + R + Y) + R = N + R.
Green + Purple = (B + Y) + (B + R) = (B + R + Y) + B = N + B.

In these three mixtures of the binaries, we have respectively the three original colours, Y, R, B, combined with N. And thus, if N be white light a restoration of the three original primaries is effected, but if (N) represent grey, obtained by mixing material colours, then the three primaries will appear dull or "impure." This dull appearance always results from the mixture of any two material colours, however brilliant or "pure" they may naturally be. These remarks are perhaps sufficient to show that all brilliant or "pure" colours may be considered equally as primaries or binaries, combined with a greater or less proportion of white light; whilst all dull or "impure" colours result from mixing pure colours with grey. In order to obtain any truly brilliant tint we must procure our colour from some natural substance and not form it by admixture. Such pure colours are comparatively rare in nature, and even those which approach the nearest to brilliancy generally contain more or less grey. Although it is particularly difficult to obtain either of the three colours which we have adopted as our primaries perfectly pure from admixture with one of the other two, we may state our theory and then we must practically contrive to make as close an approximation to such a scheme as the nature of the case will admit.

It will be evident, that any pure colour in nature, when reduced to the same shade as those in our scale (*fig.* 154.), will either exactly coincide with one of the twelve basial colours or lie between two which are contiguous. Thus a colour whose composition is 5 B + 3 Y, lies between (B + Y) and (2 B + Y), and its exact position may be ascertained, by forming fresh combinations between these two colours and their resultants as before described. Thus,

Since (2 B + Y) and (B + Y) are contiguous in the third circle,
So will (2 B + Y) — (3 B + 2 Y) — (B + Y) be in the fourth,
And (2 B + Y) — (5 B + 3 Y) — (3 B + 2 Y) — (4 B + 3 Y) — (B + Y) in the fifth, &c.

This colour therefore is one of forty-eight pure colours which would compose a fifth circle constructed on the plan alluded to. We may remark that any two colours arranged in opposite compartments added together make white or grey, and are hence styled complementary colours. Thus (2 B + Y) is exactly opposite to (2 R + Y), and these added together make up (2 B + 2 R + 2 Y) or 2 N; and so of any others.

(185.) *Impure Colours.* — From what we have said it appears, that every tertiary or other compound among material colours, that is to say every dull or "impure" colour, must be some pure colour mixed with a greater or less proportion of grey. Thus, a colour composed of (9 B + 7 Y + 4 R) is the same as (4 B + 4 Y + 4 R) + (5 B + 3 Y), which is the same as (4 N) + (5 B + 3 Y), or a combination of grey (4 N) with the pure colour represented by (5 B + 3 Y) which is one of the bluish greens. Many ternary compounds have obtained specific names; thus the different "browns" result from various proportions of grey mixed with some pure colour of which red is a constituent part; and the "Olives" are some of the greens similarly rendered impure.

In order to conceive how every possible impure colour may be formed by combining the pure colours with grey, we may take the deepest shades of all the former and having placed them in the compartments of a circle divided as before, combine them with all the shades of grey beginning with the palest in the centre and proceeding to the darkest in the circumference; and then in another circle concentric with the former, combine every shade of all the brilliant colours with the deepest shade of grey. This double arrangement gives us every possible mixture between the basial colours and grey; that is to say every possible ternary compound or impure colour. Thus in the annexed figure (155.), if the deepest shade of blue extends from (*a*) to (*b*), and the deepest shade of grey from (*b*) to (*c*), then all the shades shades of grey may be added, increasing in their intensity from (*a*) to (*b*), and all those of blue from (*b*) to (*c*), and the required results will be obtained for this single basial colour. The impure colours thus formed will also be of their deepest shades.

As we have assumed twelve pure colours out of the innumerable sets which might be formed so we may assume two impure colours corresponding to each of our basial colours, as sufficient for representing the tertiary compounds. Those may be selected which lie exactly intermediate between (*a*) and (*b*), and (*b*) and (*c*) (*fig.* 155.). The former will evidently contain a double proportion of a pure colour mixed with one of grey; and the latter a double proportion of grey mixed with one of pure colour. Thus we shall have one set of "impure" and another of "very impure" colours.

(186.) *Chromatometer.* — It will be seen that we have considered the construction of twelve "pure" colours, twelve "impure" colours, and twelve "very impure" colours to be sufficient for our scheme. But we may further adopt three separate shades of each of these thirty-six colours, to which we may also refer the shades of all natural colours; and this gives us 108 different shades. If to these we add three corresponding shades of grey we shall have in all 111 to complete

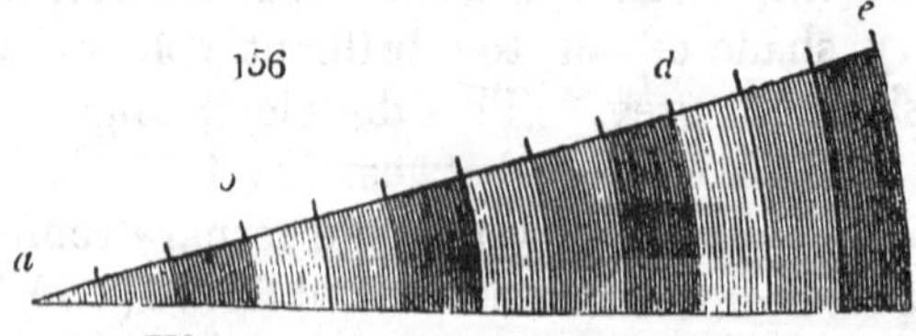

he scheme. These may be arranged in a diagram termed a "Chromatometer," which will serve for purposes of immediate reference whenever we wish to describe any colour. The annexed figure (156.) may be taken as a representation of one of its sectors, containing the three shades of grey (*a b*), and those of the "very impure" (*b c*), "impure" (*c d*), and "pure" (*d e*) blues. If the other eleven basial colours were similarly disposed round the same centre the chromatometer would be complete.

It seems unnecessary to include in this scale the different *tinges* commonly ascribed to white, black, and grey; as these after all are only very faint or dark shades of some defined colour, and may be recognised by comparison with the nearest shades expressed in the chromatometer.

(187.) *Limitation of Colour.* — It has often been observed by horticulturists, that among different varieties of the same species a limited number of colours is found, among which are not more than two out of three of the basial colours similarly disposed upon the chromatometer. Thus there are blue and red hyacinths, but none that are pure yellow; there are yellow and red dahlias, but none that are blue. The rule is not free from exceptions, still less does it apply to those flowers which have different bands of colour on their corolla. It has been conjectured that those colours which pass from green through yellow to red arise from combinations of oxygen with the chromule in its green or neutral state; whilst those which pass from green through blue to red contain a less proportion of oxygen than the green chromule itself. But as these two series meet in the same colours at both ends of such a scale it is not easy to understand how this can be the case, since the red would equally result from a union of the chromule with a maximum and with a minimum of oxygen.

(188.) *Results of Vegetable Respiration.* — From what has been said it seems necessary to conclude that carbon, in order to be fixed in vegetation must be presented to a plant in the form of carbonic acid; and that the decomposition of this gas by the direct operation of the vital principle furnishes the first step towards the organisation of brute matter.

The ultimate effects of vegetable respiration being the reverse of those which result from the analogous function in animals, have been often regarded as a remarkable provision against the gradual deterioration of our atmosphere. But the effects produced by the respiration of animals, by combustion, and by various other processes by which carbonic acid is added to the atmosphere, are of too trifling a description to enable us to appreciate their consequences under the lapse of many ages. The continued spontaneous decomposition of a large portion of dead vegetable matter, is also perpetually counteracting some portion of the beneficial effects which the fixation of carbon by plants might produce. Still it is evident that every particle of carbon in living vegetables, and likewise all that exists in those fossil bodies, coal, jet, &c. which are the altered remains of primæval vegetation, must have resulted from the decomposition of carbonic acid whose oxygen has been set free during the process of vegetable respiration. To this we may also add whatever carbon is found in animals, since this has been derived from their food primarily obtained from the vegetable kingdom. We should possess something like a measure of the extent to which vegetation has been active in altering the state of our atmosphere, if we could obtain an estimate of how much oxygen would be required to convert into carbonic acid all the carbon now fixed in organised beings, recent and fossil; and hence we might ascertain whether the atmosphere thus modified would still be fitted for our respiration or not. But in other respects there can be no doubt of the important results to which the respiration of vegetables gives rise. It is this process which prepares the organisable materials from whose subsequent elaboration are derived those infinitely varied conditions of organised matter which are essential to the development of the numerous tribes of plants which gladden the fair face of nature, and serve to nourish the myriads of animated beings which people the earth, the ocean, and the atmosphere. And lastly and most incomprehensibly, from these same materials are constructed those organised substances which seem to stand as portals to the intellectual and spiritual world — channels of direct communication by which reason and revelation may tell the frail tenants of a few mouldering atoms, of that more glorious condition which will as certainly be their heritage hereafter as their hopes and yearnings after immortality are within the actual experience of their present state.

CHAP. III.

FUNCTION OF NUTRITION CONTINUED — *Periods* 5, 6.

DIFFUSION OF PROPER JUICE (189.). — INTERCELLULAR ROTATION (193.). — LOCAL CIRCULATIONS (195.). — VEGETABLE SECRETIONS (196.). — FECULA, SUGAR, LIGNINE (197.). — PROPER JUICES (202.) — TASTE AND SCENT (210.). — EXCRETIONS (212.). — ROTATIONS OF CROPS (218.). — EXTRANEOUS DEPOSITS (219.).

FIFTH PERIOD OF NUTRITION.

(189.) *Diffusion of proper Juice.*— The crude sap having been subjected to the action of the atmosphere and the carbonic acid decomposed, the result is termed the "proper juice" or elaborated sap of the plant. This liquid has now to find its way back again into the system for the purpose of nourishing and developing the various parts. There are three distinct kinds of movement to which the proper juices of plants are subjected. The first of these is its descent and transfusion; the second is a very singular rotation of the juices contained in the vesicles and short tubes of some plants; and the third is a sort of actual though local circulation more nearly resembling the circulation of blood in animals. We propose to describe each of these under the present period, though certainly they can hardly be all considered as subordinate processes of the same function.

(190.) *Descent of Sap.*—When a ring of bark is removed from a stem or branch of a dicotyledonous plant a tumour is formed at the upper edge of the ring, which indicates a stoppage to have taken place in the descent of the elaborated sap. This stoppage by causing an excess of nutriment to accumulate above the ring, operates in improving the size and quality of fruits, and will even occasion a tree to flower and produce fruit when it would otherwise have developed nothing but leaves. No increase or at most a very slight one takes place in the diameter of the trunk below the ring; but the part above it is more developed than it otherwise would have been. If a potato be ringed in this way the buds in the axillæ of its leaves are developed in the form of little tubers, whilst none are produced on the underground stems or rhizomata. Similar effects are produced by a tight ligature; and most persons have observed the appearance which a woodbine causes on the branches of trees by twining round them. A spiral protuberance is formed immediately above and below the stricture, but more especially above it, and in process of time these swellings often become so large as to meet completely over the woodbine and embed it in the substance of the tree. The parts which lie above a ring or ligature become specifically heavier than those which are below it as Mr. Knight found in the oak, the wood above having a specific gravity of 1·14, and that below only 1·11. All these facts seem to indicate that the chief passage of the descending sap is down the bark, and towards the surface of the stem. It was supposed by some persons that an important advantage might be taken of this circumstance; and that by stripping a tree of its bark some time before it was felled, the sap would be forced to descend along the newly formed wood and thus ripen or harden it more speedily than would have been the case in the natural course of things. But experience has shown that such timber is very brittle and unfit for the purposes of building.

(191.) *Progression of the Sap.*—Although the proper juice appears to descend more especially by the bark and those portions of the tree which are towards the surface, and which are in fact the parts where the vitality of the trunk resides, there still appears to be a very general diffusion of the nutritious juice continually taking place throughout all parts of the tree, sometimes in one direction and sometimes in another. This may be shown by a contrivance of M. Biot (*fig.* 157.). A wooden wedge boiled in wax and oil to render it impervious to moisture, has a groove cut in the upper part, and is then driven into a cavity which it exactly fits in the trunk of a tree; a space is hollowed out both above and below this wedge; the roof of the cavity above it shelves towards the middle, so that the descending sap collects there and drops into the open extremity of a pipe placed in the groove to receive it. The ascending sap rises into the lower cavity which is also cut into a groove, and it is there received into another pipe placed in the bottom. In this manner a flow of sap is obtained either simultaneously from both pipes, or at separate times and in different proportions according to the state of the atmosphere, season of the year, and other circumstances

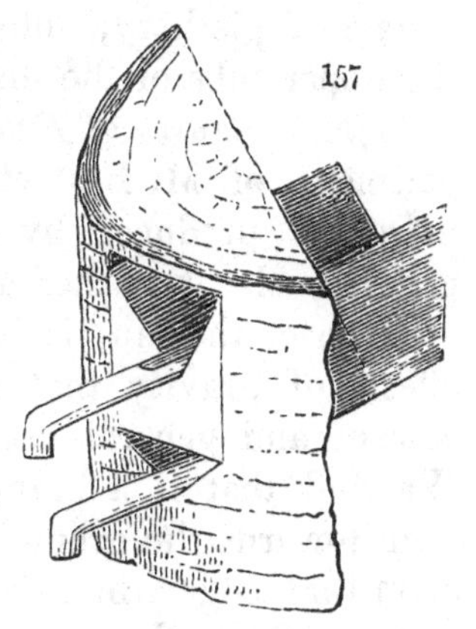
157

which influence the flow. It is observed that the descending current is generally denser and more saccharine than the ascending, although the reverse is occasionally the case after violent rains. Light appears to be the principal agent in modifying the conditions of the flow. Mild weather promotes the ascent, and a sudden cold succeeding causes a rapid descent by contracting the trunk of the tree. If the cold continue and the ground become frozen, the sap is again forced to ascend. When a thaw succeeds a frost the exhausted roots are to be replenished, and the downward current is re-established. The rapid ascent which commences in spring when the buds are beginning to burst, ceases as soon as the leaves are completely expanded. After midsummer the power of the solar rays being less energetic, and the deposition of earthy particles having obstructed the vessels of the leaf less sap is exhaled from them and the tree attains a state of plethory, indicated by an increasing flow at the upper tube of the instrument.

(192.) *Causes of Progression.* — Although these experiments of M. Biot clearly indicate that there is an influence produced by a change of temperature and probably also by other atmospheric causes on the progression of the sap, it is neither to these nor yet to the effects of gravity that we must entirely attribute the descent and general diffusion of the nutritious juices. We find that if a branch is ringed and its extremity bent towards the ground, the tumour now is produced upon that edge which is the lowest in position though furthest from the root, and consequently the returning sap has been compelled to rise into the pendent branch. Its progression is decidedly facilitated by mechanical causes, such as the wind continually agitating the stem and branches. Mr. Knight confined the stem of a tree so that it could vibrate only in one plane; and at the end of some years he observed that its section was an ellipse, whose greater axis lay in this plane.

(193.) *Intercellular Rotation.* — In the ascent,

descent and general transfusion of the sap, we can trace the operation of physical causes modifying and controlling to a considerable extent, if indeed they do not originate and entirely regulate these movements. We have now to describe a more remarkable movement of the juices of some plants, which more decidedly evinces a vital action. This movement consists in a constant rotation of the fluid contained in their vesicles and tubes, and rendered apparent by the presence of minute globules of vegetable matter floating in it. The original disovery of this phenomenon was made about a century ago by Corti, who first observed it in the *Caulinia fragilis,* a maritime plant found on the shores of Italy. His observations appear to have been generally neglected until lately, when the re-discovery of the phenomenon in other plants has excited the attention of botanists. It may readily be seen with a good lens in Valisneria, Hydrocharis, Potamogeton, and other aquatic genera, but more especially in the genus Chara. It has also been observed in the terrestrial genera Cucurbita, Cucumis, Pistia, and others; and is more especially observable in the hairs of many species. It appears to be a universal property of the cellular tissue though it is impossible in many cases to detect it, either on account of the want of sufficient transparency in the membrane or from the absence of the granular matter by whose presence alone the rotation of the fluid itself can be observed. We shall explain the phenomenon as it may be seen in the Chara with a lens of about the tenth of an inch focal distance or even of less power.

(194.) *Rotation of Fluid in Chara.* — This genus may be divided into two sections, which are considered as distinct genera by Agardh. In one of them, the true Chara, the stems are composed of a central tube jointed at intervals and surrounded by a row of smaller tubes. In the other section, or genus Nitella, the stems consist of single tubes jointed as before. If we select a species of the first section it will be necessary to clear away the outer tubes which are always more or less encrusted with carbonate of lime, in order to expose the inner tube in which the rotation of the fluid may be seen. This is an operation requiring some little delicacy; and the choice of a species of the other section (*Nitella*) is to be preferred, in which the tubes are generally very transparent and require no preliminary preparation to clean their surface. At the joints of the stem are whorls of branches (*fig.* 158.) composed also of short tubes, in each of which the same rotation of the contained fluid may be seen. If an entire tube occupying the space between two joints be detached and placed under the microscope, its inner surface appears to be studded with minute green granules arranged in lines, which do not run parallel to the axis of the tube but wind in a spiral direction from one extremity to the other. They are studded over the whole of the interior, with the exception of two narrow spaces on opposite sides of the tube forming two spiral lines from end to end. The globules of transparent gelatinous matter dispersed through the fluid are in constant motion, being directed by a current up one side of the tube and back again by the other. The course of this current is regulated by the spiral arrangement of the granules, and it moves in opposite directions on contrary sides of the clear spaces on the inner surface of the tube. The rotation continues in a detached portion, for several days; and if the tube is tied at intervals between the joints the fluid between two ligatures still continues to circulate, even though the extremities of the tube should be cut away. The motion here described is precisely similar to what takes place in the tubes of Corallines, and must unquestionably be considered as the result of a vital action.

158

(195.) *Local Circulations.* — It was in the year 1820, that a distinguished naturalist, M. Schultes, first announced his discovery of a peculiar movement in the juices of plants, which more nearly resembles the circulation of the blood in animals than any thing which had formerly been observed. The existence of such a circulation had been strongly suspected before; but as the experiments upon which his actual detection of the phenomenon depended were difficult to verify, his account was much disputed until recently when he obtained the prize which the Academy of Sciences at Paris had proposed for the purpose of eliciting further investigations on the subject. His memoir has not, hitherto we believe made its appearance; but the committee appointed to examine its merits have made a favourable report of its contents published in the "Archives de Botanique" for 1833; and from this and a former paper in the "Annales des Sciences," we have gleaned the following particulars: — The liquid, whose movement is described and which M. Schultes terms the "latex," is sometimes transparent and colourless but in many cases opaque, and either milk-white, yellow, red, orange, or brown. The colours depend upon the presence of innumerable minute globules which are constantly agitated as if by a spontaneous motion, and appear to be alternately attracted and repelled by each other. This liquid is considered to be the proper juice of the plant secreted from the crude sap in the intercellular passages and consequently analogous to the blood of animals as was long since suggested by Grew, who further likened the lymphatic or crude sap to their chyle. It is contained in delicate transparent membranous tubes, which become cylindrical when isolated, but when packed together in bundles assume a polygonal shape. In young shoots it is difficult to detect them, on account of their extreme transparency and tenuity; but they may be extracted with considerable facility from older parts. They have been observed very generally in Monocotyledons and in Dicotyledons, excepting in the few species in which no tracheæ have been hitherto noticed. They frequently intercommunicate or anastomose by means of lateral branches, and sometimes form a regular network (see art. 27. *fig.* 15.). They occur in the woody fibre, in the bark, occasionally even in the pith, and very frequently surround the tracheæ. They exist in greatest complexity in the root, from whence they proceed in parallel lines up the stem into the leaves and flowers and then return again to the root, the ascending and descending branches anastomosing throughout their course. The movement of the latex can be witnessed only in those parts which happen to be very transparent; and it has not been actually seen in many plants. The *Ficus elastica*, *Chelidonium majus*, and *Alisma plantago*, are the species upon which most of the observations hitherto recorded have been made. Distinct currents are observed traversing the vital vessels, and passing through the lateral connecting tubes or branches into the principal channels. These currents follow no one determinate course, but are very inconstant in their direction—some proceeding up and others down, some to the right and others to the left; the motion occasionally stopping suddenly, and then recommencing. In detached fragments of the plant it will continue from five minutes to half an hour, according to circumstances; but M. Schultes has been able so to adjust his lens as to witness the flow in the growing plant. The action is suddenly checked by cold, and again recommences with an elevation of temperature. The effect does not seem to depend upon a contractile power of the tubes, because the latex flows chiefly or entirely from one end of a tube even when it has an orifice open at both extremities. The appearance is very similar to the circulation of the blood in the fœtus contained in a bird's egg before the heart is formed; but is more especially analogous to the circulation of some of the lowest tribes of animals, as in the Diplozoon paradoxum, which may be divided into two parts and the blood still continue to circulate for three or four hours in each. By a strong electric shock, the force by which the latex is propelled is paralysed, and its motion arrested.

SIXTH PERIOD OF NUTRITION.

(196.) *Vegetable Secretions.*—In describing the process by which we have supposed the first step to be made towards the organisation of those materials which enter into the vegetable structure, we have considered gum to be the immediate result of the fixation of carbon in combination with the two elements of water; and that this substance is formed by all those parts of plants which almost universally acquire a green tinge. We further stated that there were three other substances nearly allied to gum in chemical composition, which might also be considered as destined for the nourishment of the plant. It is probable that these substances are only slight modifications of gum, produced by its subsequent elaboration in the cellular tissue. It is impossible, however, to point out the specific organs which are appropriated to this office. In some cases a distinct glandular structure is very apparent, and the immediate secretions effected by it are collected in an isolated form; but in others there is no apparent difference between the organisation of those parts in which the secretions are produced and the rest of the cellular tissue.

(197.) *Fecula.*— The first of the three alimentary products which we shall further notice is fecula. This substance forms minute spheroidal grains in the cellular tissue, and must be considered rather as a distinctly organised product than as a secreted matter. Each grain consists of an insoluble pellicle or integument, containing a soluble substance which seems to be pure gum, or some material scarcely differing from it in any essential character. These grains are not altered by the action of alcohol, ether, or cold water; but in hot water the pellicle bursts, the contained matter exudes, and the whole mass becomes a paste. The specific gravity of fecula is about 1·53. It may be obtained from the pulp of fruits, tubers, succulent stems, and other parts of various plants. That which is derived from corn and the potato is familiarly termed starch. Sago (from the stems of a palm), tapioca (from the tubers of the Jatropha manihot), arrow-root (from the rhizomata of the Maranta arundinacea), are all so many varieties of fecula. This substance is highly alimentary and is largely stored up in various parts of vegetables where it forms magazines of nutriment, apparently destined for the future development of the buds and ripening of the seed. It is a material of all others the most important as an article of human food, and is providentially provided for our use in the greatest abundance. It bears a striking analogy to the fat of animals, even in the general structure of its component parts according to some, but more evidently in the uses to which it is subservient in the economy of vegetation. The formation and subsequent re-absorption of fecula is rendered very evident, by comparing the different quantities found in plants of the same species at different seasons of the year. The following table shows us the gradual accumulation which takes place in 100 pounds of potatoes between August and November, and the subsequent diminution from March to May:—

Aug.	*Sept.*	*Oct.*	*Nov.*	*March.*	*April.*	*May.*
10	14½	14¾	17	17	13¾	10

(198.) *Plants containing Fecula.*—The following list contains a few of the principal plants which furnish fecula in the greatest abundance, and the figures give the percentage yielded by the several organs from which it is extracted. These numbers may also be considered to a certain extent indicative of the degrees of nourishment which each is capable of affording:—

Maize - - -	80 to 92	seed.
Rice - - -	80 to 85	
Wheat - - -	70 to 77	
Rye - - -	61	
Oats - - -	59	
Peas - - -	50	
French beans - - -	46	
Kidney beans - - -	34	
Lentils - - - -	32	
Amomum curcuma -	26	rhizoma.
Dioscorea triloba - -	25	
Potato - - - -	24	tuber.
Tapioca (*Jatropha manihot*) -	13·5	root.
Sweet Potato (*Ipomœa batatas*)	13·3	
Arrow-root (*Maranta arundinacea*)	12·5	rhizoma.
Canna coccinea - -	12·5	
Breadfruit (*Artocarpus incisa*) -	3·2	fruit.

(199.) *Sugar.* — There are numerous modifications of sugar, all of which may be referred to two general heads. The one class, as the sugars of the sugar-cane and beet-root, contains a less proportion of water in combination with an equal quantity of carbon than the other class, which includes the sugars extracted from raisins, manna, &c. Some are crystallisable others not. The purest obtained from the sugar-cane has a specific gravity of 1·605, and is composed of about 42 per cent. of carbon and 58 of water. In the East Indies the canes yield about 17 per cent., and in America 14 per cent. of sugar; but in our hot-houses they produce scarcely any. All sugars are readily soluble in water but less so in alcohol, into which latter fluid they may themselves be converted by the process of fermentation; thus the quantity of ardent spirits which may be extracted from any vegetable is in proportion to the sugar it contains. This substance bears a striking affinity to gum in its chemical composition, and is very commonly dissolved in the juices of plants. After it has been formed it is again very easily altered during the progress of vegetation; a fact of considerable importance to the cultivator, who must be cautious to collect the produce of his canes at the season when the sugar is most abundantly generated and before it sustains such alteration. The flowering of the cane exhausts the sugar in the stem; and that which is so abundantly contained in the cortical system of the root of the beet is ultimately carried into the upper parts of the plant, and similarly exhausted during its inflorescence.

(200.) *Lignine.* — This substance is contained in the elongated vesicles termed closters (art. 16. *fig.* 3. *c*), of which the woody fibre is composed. It does not appear that it has ever been submitted to a careful analysis, or accurately examined in a detached form. After various matters have been abstracted from the woody fibre, such as certain salts, gummy particles, and others, there then remains about 96 per cent. of an insoluble substance, composed of nearly equal proportions of water and carbon. But this is a compound material, consisting both of the thin pellicle which formed the vesicles themselves as well as of the lignine which they contained. The resemblance which lignine bears to gum is not so striking as in the case of the two materials just described, nor does it appear to answer any ulterior purpose of nutrition after it has become secreted; but it remains unchanged in the cells, and imparts to wood the varied qualities and colours which different species present. Its specific gravity varies being 1·459 in the maple, and 1·534 in the oak.

(201.) *Vegetable Products.* — Besides the four materials gum, fecula, sugar, and lignine, which we consider as the simplest modifications which the nutritious and organisable materials found in the vegetable structure can assume, there is an interminable catalogue of other substances which may be extracted from the juices of different plants, all of which have been formed by secretion in some part or other of their structure.

Some are the results of disease, whilst others are more abundantly formed when the plants which produce them are placed in peculiar soils and situations. Some occur in a very few species only, whilst others are characteristic of whole families. None of them are so abundantly diffused as the four nutritive substances already described; and they all materially differ from these, by having either the oxygen or the hydrogen which they contain in greater excess than would be necessary to form water. These may therefore be termed hyperoxygenated and hyperhydrogenated products, when contrasted with the others. Little is at present known of the exact manner in which these various products are formed. Their complete enumeration belongs to the department of chemical Botany; and we can here pretend to do no more than point out some of the principal groups, and mention a few of their most striking peculiarities.

(202.) *Proper Juices.* — Several of the products elaborated in the leaves and cortical parts, are dissolved in those proper juices of plants which in art. 195. we described as the latex or vital fluid, analogous to the blood of animals. But as these juices are very different in their characters in different species, as they are not clearly defined in some and above all as they act as poisons when imbibed by the roots, De Candolle imagines that they ought more properly to be considered as secretions of a recrementitial nature, analogous to the bile and others in the animal economy. Some of these products even contain azote, and by this circumstance are brought into closer resemblance with animal matter. The more remarkable materials found in the proper juices of plants are milks, resins, and oils.

(203.) *Milks.* — These are generally of an opaque white, though some are variously coloured. They abound in many species, and are highly characteristic of certain natural families, as the Euphorbiaceæ,

Apocyneæ, Artocarpeæ, &c. They differ very remarkably in their characters; for although a large portion are noxious, and even highly poisonous, some on the contrary are wholesome and nutritious. There are several substances found in the composition of these milks, of which we may mention the following: —

1. *Caoutchouc,* or Indian rubber is abundant in some of them, and may be readily obtained from several trees of different families growing in tropical climates. All that is requisite for the purpose of procuring this material, is to receive the milk into suitable vessels as it flows from a wound in the bark and to allow its aqueous particles to evaporate, when the caoutchouc remains in a solid form.

2. *Opium* is procured by inspissating the milk of the poppy, and is also found in other plants.

3. *The Cow-Tree.* — One of the most remarkable phenomena of the vegetable world is the cow-tree described by Humboldt in the following terms, as growing in the Cordilleras of South America: — "On the barren flank of a rock grows a tree with dry and leather-like leaves; its large woody roots can scarcely penetrate into the stony soil. For several months in the year not a single shower moistens its foliage. Its branches appear dead and dried; yet as soon as the trunk is pierced, there flows from it a sweet and nourishing milk. It is at sunrise that this vegetable fountain is most abundant. The natives are then to be seen hastening from all quarters, furnished with large bowls to receive the milk, which grows yellow and thickens at the surface. Some empty their bowls under the tree, while others carry home the juice to their children. The milk obtained by incisions made in the trunk is glutinous, tolerably thick, free from all acrimony, and of an agreeable and balmy smell. It was offered to us in the shell of the tutuno, or calabash tree. We drank a considerable quantity of it in the evening, before we went to bed, and very early in the morning, without experiencing the slightest injurious

effect. The viscosity of the milk alone renders it somewhat disagreeable. The negroes and free labourers drink it, dipping into it their maize, or cassava bread." Mr. Lockhart has subsequently afforded the following additional particulars concerning this tree:—"The *Palo de vaca* is a tree of large dimensions. The one that I procured the juice from had a trunk seven feet in diameter, and it was one hundred feet from the root to the first branch. The milk was obtained by making a spiral incision into the bark. The milk is used by the inhabitants wherever it is known. I drank a pint of it without experiencing the least inconvenience. In taste and consistence it much resembles sweet cream, and possesses an agreeable smell."

(204.) *Receptacles for Milk.*—All the various milky juices reside in the bark and leaves, and are not found in the wood. They are contained in distinct receptacles, and may be extracted by means of incisions chiefly in the upper parts of plants, and which do not extend deeper than the bark; otherwise they would be diluted and impoverished by mixing with the ascending sap. M. Bertholet has recorded a remarkable instance of the harmless quality of the sap in the interior of a plant, whose bark is filled with a milky proper juice of a poisonous nature. He describes the natives of Teneriffe as being in the habit of removing the bark from the *Euphorbia canariensis*, and then sucking the inner portion of the stem in order to quench their thirst, this part containing a considerable quantity of limpid and non-elaborated sap. The reservoirs which contain the milky juice of the wild lettuce (*Lactuca virosa*) are so remarkably irritable that the slightest touch is sufficient to cause it to be ejected from them with considerable force. When this plant is about to flower, if an insect happens to crawl over the surface of the stalk any where near its summit a jet of milk is propelled. In general plants which secrete these milky juices love the light; few are found to affect shady situations, and none are aquatics. By cultivation, their noxious properties may be greatly subdued.

(205.) *Resins.* — This class contains certain substances separated from the proper juice by some process of secretion; and not having any peculiar channels appropriated to their reception, they form cavities and force passages for themselves in the cellular tissue. Occasionally they exude from the surface of the stem; but this must be considered accidental and not the result of any provision made for their excretion, as is the case with some substances which exude from certain glands on the surface.

(206.) *Oils.* — There are two classes of oils secreted by plants: the one contains the highly volatile or essential oils as they are termed, which impart the fragrant or disagreeable odours peculiar to different plants; and the other the fixed oils, such as those extracted from the fruit of the olive, the seeds of flax, &c.

(207.) *Volatile Oils.* — The first kind are generally contained in spherical or oblong cells in the leaves and cortical parts of plants; when held to the light these parts appear as if they were punctured, owing to the superior transparency of the receptacles in which the oil is deposited. The St. John's-wort (*Hypericum perforatum*) and any of the myrtle tribe are familiar examples of this fact. In the Umbelliferæ the oil accumulates in oblong club-shaped receptacles, termed "vittæ," which are placed between the coats of the seed-vessel; and it is remarkable that their number and general appearance is so constantly the same for each separate species that important generic characters are derived from this circumstance.

(208.) *Camphor* is deposited upon the evaporation of certain volatile oils, especially those extracted from some of the Labiatæ, as the common lavender.

(209.) *Fixed Oils.* — These are rarely found in the cortical parts like the others, but are for the most part extracted from the seed or its envelopes, and sometimes from the pericarp, as in the olive. In these cases they are readily convertible by some natural process into a nutritious emulsion; and then appear to be destined to feed the young plant during the early stages of its development.

The following table shows the percentage of fixed oil obtained from the seeds of a few plants:—

Nut - -	60
Cress - -	58
Walnut - -	50
Poppy - -	47
Almond - -	46

(210.) *Taste and Scent of Plants.*— It will readily be conceived that the peculiar tastes and odours met with in different species, must depend entirely upon the nature of the various matters which are secreted by them. Attempts have been made to classify the various impressions which are thus made upon the sensorium, and odours have been arranged into classes, under the terms aromatic, fœtid, acrid, alliaceous, musky, &c. Such classifications at the best are highly empyrical, and any arrangement which could be founded on an accurate knowledge of the chemical nature of these substances would be far preferable; but our extreme ignorance on these points will not justify the attempt at present. The delicate perfumes emitted by certain flowers, as well as the more powerful and often disagreeable scents afforded by the herbage of some plants, generally depend upon the diffusion of a volatile oil. In some cases this oil is magazined in the stalks and leaves, and is rendered more sensible the more these parts are rubbed or bruised. In the flower especially, the oily particles which produce the odour seem to be diffused as fast as they are secreted; and hence it happens that the greater number of plants are more powerfully scented at one particular part of the day and that almost all flowers are most fragrant towards night. There are some, specially termed "night-scented," which are extremely powerful after sunset though they emit little or no odour by day; and several of these as the night-scented stock, geranium, wallflower, gladiolus, &c., are further remarkable from possessing a peculiar brown and lurid tint. The flowers of the splendid *Cereus grandiflorus* begin to expand about seven o'clock in the evening, attain their full beauty and put forth their powerfully fragrant odour before midnight, and are completely faded before sunrise. Some of the singular tribe of Stapelias are disgustingly nauseous in the scent which they emit, strongly resembling the most offensive carrion; so much so indeed that even flies and other carnivorous insects are deceived by the similarity, and very frequently deposit their eggs in their blossom.

(211.) *Impressions made by Odours.* — The scents emitted by certain flowers make very different impressions upon the nerves of different people; and some persons can readily perceive a powerful odour where others are nearly or entirely insensible to its impression, although they may not be defective in other instances in the sense of smelling. Very deleterious impressions are made on some constitutions by the odours of strong-scented flowers. The most dangerous symptoms have occurred in persons especially females with weak nerves, merely by their remaining in a room where certain flowers have been placed; and even violets are not exempt from a bad reputation. Instances of death have been recorded which were considered to have been occasioned by effects of this kind; and Linnæus mentions a case where the odour from the Rose-bay (*Nereum oleander*) was supposed to have proved fatal to the constitution of one person. Prussic acid may be instanced as abounding in the leaves of the common laurel (*Prunus lauroceratus*) to so great an extent, that if one of them be cut into small pieces and placed under a wine-glass, and a wasp or other insect be introduced under the glass it will be completely stupefied in two minutes.

(212.) *Excretions.* — We have still to allude to

a class of substances which are *excreted* from plants by various glands seated on the surface of their stems, leaves, and other organs. Many of them are of the same description as those which are formed within the plant by internal secretions, such as acids, oils, &c.; but some of them are peculiar. They may be considered as more strictly analogous to the various excrementitious matters ejected by animals than those of the former class; and the glands by which they are formed are for the most part more complex and better defined than those which are seated in the interior of plants. The external glands (see art. 31. and *fig.* 20.) by which these matters are excreted often form a sort of clammy pubescence upon the epidermis. They frequently resemble hairs tipped with a little globular mass by which the excreted matter is more especially elaborated.

(213.) *Fraxinella.* — The common Fraxinella is covered with minute glands which excrete a volatile oil. This is continually evaporating from its surface, and on a calm still evening forms a highly inflammable atmosphere round the plant. If a candle be brought near it, the plant is enveloped by a transient flame without sustaining any injury from the experiment.

(214.) *Stings.* — The stinging plants prepare a caustic juice which is contained in a cellular bag surmounted by a hollow bristle. When the bristle is gently pressed the fluid is forced through it and flows out at the summit through a minute orifice, as we have stated (art. 31. and *fig.* 20. *a*). If the bristle enters a pore of the skin, the caustic fluid is introduced and produces the painful sensations familiar to all who have ever handled a common nettle. The Loasæ have stings which give a still more irritating sensation than the nettles. The Malpighiæ are furnished with a multitude of doubly pointed bristles which lie parallel to the surface of their leaves, to which they are attached by a short hollow stem. These contain a slightly caustic fluid.

(215.) *Glue.*—The gummy excretions on the stems of

certain plants, as the fly-catching Lychnises (*Lychnis armeria* and others) appear to be composed of a material of the same nature as common birdlime extracted from the bark of the holly. Several kind of leaf-buds, as those of the horse-chestnut, are coated over with a glutinous insoluble excretion apparently intended to secure them from the ill effects of moisture.

(216.) *Wax*—is a very abundant excretion from many plants. It forms a delicate powder on the surface of certain fruits, as the substance termed the "bloom" on the plum. It is so plentiful on the surface of poplar leaves, that a manufactory was at one time established in Italy for the purpose of procuring it from them as a material for commerce. It is very abundantly furnished by some palms in tropical countries, where it is advantageously employed for economical purposes; but the *Myrica cerifera* is the plant which affords it in the greatest abundance. Its fruit is completely enveloped in a coat of wax, and when thrown into boiling water the wax melts and floats to the surface where it is skimmed off. It has a slightly green tinge which can be removed by chlorine, and it may then be formed into candles resembling spermaceti. This fruit yields about one ninth per cent. of its weight in wax. All the kinds of vegetable wax are closely allied to common bees' wax in several properties, though essentially distinguished from it by others.

(217.) *Radical Excretions.*—But of all excretions proceeding from plants, some of the least-known are perhaps the most important in an economical point of view. It was not until very recently that their properties had been made a subject of experimental inquiry, or even that their existence had been clearly established; but the partial results hitherto obtained have opened a wide field for speculation. The excretions to which we allude are discharged from the root, and may be detected by a very simple experiment. If young French beans, for example, be placed in a glass containing distilled water, at the end of a few days this water will be found strongly impregnated by a matter excreted from the roots. A fresh plant should be placed daily in the water, to avoid the effects which might otherwise be produced by an incipient decomposition. It is also found that the matters thus procured from plants of different families are dissimilar. Thus that which is excreted by the Leguminosæ contains an abundance of mucilage, whilst that which exudes from the Gramineæ has very little. The Chicoraceæ excrete a bitter matter analogous to opium; the Euphorbiaceæ a gum-resinous matter, &c.

(218.) *Rotation of Crops.* — So far as observations have hitherto been made, it appears probable that the excretions given out by plants of different families possess very different qualities, and act differently upon other plants. It had been long known to gardeners that flowers and fruit-trees will not prosper so well when they have been planted in a situation where others of the same kind had previously grown, as if they were planted in situations where they succeeded to others of a different kind. It is also a well-established fact in forestry, that when a wood principally composed of one species of timber trees has been cleared, the trees which then spring up spontaneously and supply the place of the former growth are for the most part of a different species. And lastly, the agriculturist has established a rotation of crops upon experimental proof that grain of one kind succeeds better when it follows certain other kinds, than when it is sown immediately after a crop of the same plant. The various theories which had formerly been proposed to account for these facts were all liable to serious objections; but M. De Candolle has suggested the probability, that the excretions of any one plant although they may be noxious to others of the same species, genus, or family, may nevertheless be perfectly harmless or even beneficial to plants of other families. In this manner he would account for the fact, that plants of the natural order Leguminosæ (as vetches, tares, &c.), prepare or improve the soil for those of the Gramineæ (various kinds of corn, &c.). If the farmer by further experimental research should ever be able to establish an extensive series of facts of this description, he may expect to grow a succession of crops with comparatively little manure and without ever being obliged to let his land lie fallow. In the present state of this inquiry it would be idle to say much upon the possible advantages which may be expected from the confirmation of this theory; but it must be evident to the most prejudiced admirer of old customs, that we cannot expect to make any real progress in the various branches of human knowledge, agriculture among the rest, until we have obtained clearer notions and a sounder theory respecting the fundamental principles upon which the successful practice of any pursuit depends.

(219.) *Extraneous Matters.*— Besides those numerous products directly secreted by plants, and which are the immediate results of vegetable action, there are many others which have either been accidentally absorbed with the water that enters through the spongioles and pores, or else have resulted from subsequent combinations chemically effected between matters so introduced and the undoubted products of vegetation. All matters however which are accidentally introduced, form only a very slight per centage of the weight of the whole mass. They compose the various earthy, saline, metallic, and other ingredients found in the ashes of plants, after combustion has dissipated all the purely vegetable products. They generally exist in the greatest quantity in those plants, or parts of plants, where the process of exhalation has been carried on with the greatest rapidity. Hence they abound more in the leaves than in other parts, and more in the bark than in the wood. Herbaceous plants for similar reasons furnish more ashes than trees.

(220.) *Earths.*—*Lime* is the earth which is most universally present in the ashes of plants, generally in the form of a carbonate, but also in union with other mineral and vegetable acids. Carbonate of lime is largely deposited in the stems of some of the Charæ, which it completely incrusts with stony matter. — *Silica* is the earth which next to lime occurs in the greatest abundance, especially among some of the monocotyledonous tribes. The glossy surfaces of canes, reeds, and other grasses, are composed of a very large percentage of it; and if two canes be rubbed together in the dark, they emit a flash of light similar to that which is obtained by the friction of two quartz pebbles. When a stack of corn or hay has been rapidly consumed, the ashes are fused into a semi-vitrified mass: the straw abounding both with silica and an alkali, the two chief ingredients necessary to the formation of such a compound. In the hollow portions of the stem between the joints of the bamboo, a substance named tabasheer is deposited in lumps which very much resemble fragments of opaque and semitransparent opal. This remarkable deposit contains 70 per cent. of pure silica, and possesses very peculiar and curious optical properties. Silica is also deposited in little semi-crystalline lumps along the angles of the stems of some species of Equiseta, especially the *Equisetum hyemale* or Dutch reed, which from this circumstance is serviceable to watchmakers and others in polishing their work.

(221.) *Salts.* — The salts of potash are particularly abundant in most plants, but the salts of soda are more especially confined to such as grow near the sea. It is however remarkable, that plants which abound in the salts of soda whilst growing in these latter situations, secrete the salts of potash when they are no longer within the influence of the sea. In such plants, it is difficult not to believe that the presence of one or other of these alkalis is in some way beneficial to their health, even though it may not form any essential part of their structure. The common soda of commerce is a carbonate obtained from the incineration of several maritime plants and sea weeds, and is largely prepared on the shores of the Mediterranean for the European market.

(222.) *Origin of extraneous Deposits.*—The various other products, such as oxides, metallic salts, &c., which occur in small quantities in the ashes of plants, have all been either derived immediately from the soil or introduced in some way by absorption from the atmosphere. It seems clearly established that none of them ought to be considered as the direct product of any vegetative function, as was once supposed; and it has been satisfactorily shown that however carefully the experiments may have been made which favour such a theory, and however cautiously the means may have been taken for excluding all foreign matters from access to the growing plant, error was unavoidable. The extreme minuteness of the elementary organs of plants, and the more delicate *manipulations* of a natural chemistry, are capable of separating the minutest portions of foreign matters from the materials with which they are brought in contact, however carefully and accurately these materials may have been purified and cleansed by artificial processes. It seems to be impossible for instance to provide even distilled water so pure, but what some traces or other of foreign matter may be detected in it.

CHAP. IV.

FUNCTION OF NUTRITION CONTINUED — *Period* 7.

ASSIMILATION (223.). — PRUNING (225.). — GRAFTING (227.). — DEVELOPMENT (230.). — NUTRITION OF CRYPTOGAMIC PLANTS (233.). — PARASITIC PLANTS (234.). — DURATION OF LIFE (235.). — VEGETABLE INDIVIDUALS (236.). — LONGEVITY OF TREES (239.).

SEVENTH PERIOD OF NUTRITION.

(223.) *Assimilation.* — THE chief end and object of the various processes which we have been describing, is the manufacture of the materials which are ultimately to be assimilated into the vegetable structure, and by which it is to be nourished and developed in all its parts. Of the precise manner in which the assimilation of this nutriment takes place we know nothing, and the first steps towards the formation and development of any organised being are entirely concealed from us. We may indeed observe when a gradual organisation of matter is taking place; but there is no stage in the process from whence we may not refer back to some previous state, out of which it appears to have emerged imperceptibly and inexplicably; and it is utterly impossible to note with any degree of accuracy, either the precise manner or exact time when the first traces of any new condition of organisation commenced. In other words, as soon as we can distinguish an organ it already exists in a developed form, however faintly its subordinate parts may be indicated.

(224.) *Growth of the Tissues.* — In dicotyledonous trees, as we have observed (art. 34. 2.), the new tissue makes its appearance between the old wood and old bark. In the earliest stage in which it is discoverable it appears as a thick clammy fluid termed the cambium, which gradually assumes the character of a newly formed cellular tissue intermixed with vessels which are disposed longitudinally through the stem. It should seem that the cellular tissue at least is developed from the old tissue, as may be shown experimentally by grafting a branch containing a wood of one colour on a tree whose wood is of a different colour as a peach on a plum. The new wood retains the distinctive characters of the parts round which it is formed, the graft increasing by pale coloured layers and the stock by layers of a reddish colour, even though these latter have been nourished by the descending sap elaborated in the leaves of the former. Different theories have been proposed in order to account for the manner in which the cellular tissue increases. Some suppose that the young cells are developed within the old ones, which they ultimately rupture and replace; but of this there is no good evidence. Others consider the opaque dots discernible on the surface of some cells to be nascent vesicles, which are afterwards developed on the outside of the old ones; and this is a more probable hypothesis than the last. According to a third opinion, an old cell becomes separated into compartments by the formation of a transverse diaphragm, and each compartment afterwards develops into a separate cell. The formation of the fresh vessels is still more ambiguous than that of the cells. One theory considers them analogous to descending roots proceeding from the buds placed in the axillæ of the leaves, and supposes them to be continuous throughout the whole length of the longest stems. But as vessels are formed, though of small dimensions, in those parts of the stem which are below the place where a ring of bark has been removed, this supposition is untenable. It seems more probable that the vessels have a common origin with the vesicles, or are modifications of them; and that a long vessel was originally composed of several parts.

(225.) *Effects of Pruning.* — The objects to be obtained by pruning are various. The gardener employs this resource as the means of improving the general form which he wishes his ornamental shrubs to assume; and he prunes his fruit trees in order that they may bear fruit of larger size and improved flavour. With these questions we have nothing to do in this place. The results of pruning which we propose to notice are such as are produced *internally* at places where the knife has been employed, particularly for the purpose of improving the quality of timber. This is attempted by removing superfluous branches, which compels the main trunk to become a straight clean shaft. The effect of every wound of this kind is to expose a portion of the older or innermost parts of the woody layers, which are incapable of generating fresh tissue. The consequence is that such parts cannot be healed over, excepting by the growth of the newest tissue round the edge of the wound. This tissue gradually extends itself from the edges over the whole surface of the wound until the opposite sides meet, and then grafting together unite into one continuous mass: but the new wood contracts no union with the surface of the old wood exposed by the operation of pruning. As the growing tissue which coats over a wound depends upon the returning sap for its supply of nutriment, no wound produced by cutting off a branch at some distance from the main trunk can ever heal. In this case there are no leaves beyond the exposed surface to supply it with proper juice, and whatever descends from the main stem is carried into the branch, and consumed in developing the buds and tissue on the lower part of it before it can arrive at its extremity. But where the branch is lopped near the trunk and a "snag" (as it is technically termed) has been left, the descending sap flows into this stump in sufficient abundance to enable the tissue to close over the exposed extremity. As the trunk increases these snags are completely embedded and greatly injure the timber; especially as they generally become more or less rotten at the exposed extremity before the new tissue has had time to coat it over. Of all descriptions of wounds those which are the nearest to the main stem heal the quickest, and this shows us the propriety of pruning as close as possible to the trunk, whenever a branch is to be removed for the purpose of improving the timber. The new tissue increases with great rapidity chiefly from above downwards, but also from the sides of the wound, and a little likewise at the base, until it has spread over the whole surface. The extent of the injury introduced into the timber is best seen by forcibly separating the new wood from the surface over which it has spread; when the latter will always be found exactly as it was left at the time it was covered up, with the mark of the knife upon it or with any portions of decay which may afterwards have taken place. This is sometimes seen in trees upon which deep inscriptions have been carved. Wherever the letters have penetrated below the bark into the woody layers an impression is left in them; and however long the new wood may have been formed over them, they will be found beneath it whenever the outer portion is removed. Birds' nests, stags' horns, an image of the Virgin Mary, and many other articles are described as having been found in the very heart of some trees, where they were unquestionably embedded by the enlargement of the stem in the way we have described.

(226.) *Precautions to be observed in Pruning.*— From what we have stated it is evident, that wherever a branch has been pruned off a blemish is inevitably introduced; and consequently where pruning can be avoided it should never be resorted to; but where it is really necessary it should be performed as early as possible, before the branch has attained any considerable dimensions. Even rubbing off the buds should be preferred to regular pruning. The cut also should be made close to the stem, and as nearly vertical as possible; the latter precaution prevents the accumulation of water upon the surface of the wound, after the newly developed wood has formed a swollen border round its edges. If the cut is perfectly smooth it will be the sooner healed; and its surface may be protected by some compost (such as that which is known by the name of Forsyth's mixture) whenever the wound is unavoidably large. An opinion has gone abroad that it is possible to diminish the blemish which pruning necessarily occasions in timber, by lopping the extremities of a branch and causing them to die and rot off in a natural manner. Supposing it were true that a branch thus treated always did die,—which is by no means a necessary consequence,—all that could be gained by such a mode of proceeding would be the introduction of the rotten stump of the lopped branch into the heart of the tree instead of the clean scar which close pruning produces. It is not true, as some suppose, that any natural *sloughing off* of the decayed part takes place or that the old and new wood can ever completely unite together; but in all cases it will be found that the new wood has grown over the old wound, and that the surface of the latter is preserved exactly in the state in which it was embedded. The knots in deal and other timbers are defects produced by the process of "natural pruning," as it has been termed, and such defects are inevitably greater than those which result from artificial pruning performed on branches of the same dimensions and cut off close to the stem.

(227.) *Grafts.* — Every one is acquainted with the fact, that certain portions of some plants may be grafted upon others, and that the tissues of the "graft" and "stock" as the two are named will completely unite and vegetate together as though they were parts of the same individual. The effects thus artificially produced are occasionally observed to take place naturally: two branches of the same tree being sometimes found grafted together, where they have been wounded by mutual attrition. When ivy has grown to a considerable size its branches often interlace and graft together in various places, till the whole forms a rude network upon the trunk of the tree up which it has climbed. Although it is so easy for two parts of different individuals of the same species to graft together, it requires great care and precaution to secure such a union between two different species. In dicotyledonous plants the two alburnums and the two libers must be placed in contact, and then the line of junction between the two cambiums will also be complete and the newly formed tissues will readily unite. De Candolle thinks it likely, in contradiction to the common opinion, that the ascending sap being attracted by the graft will first produce a union between the two alburnums, and that the descending sap then effects the union of the two libers. The chief requisite in this operation is the near relationship of the two species; and it never succeeds excepting between such as are of the same genus or at least between allied genera of the same family. The ancients were of a very different opinion, and considered it possible to graft any two plants together. Thus Virgil: —

> "Et steriles platani malos gessere valentes,
> Castaneæ fagos, ornusque incanuit albo
> Flore Tyri, glandemque sues fregere sub ulmis."

Pliny has recorded a marvellous instance of a grafted tree bearing a variety of different fruits, which he tells us he himself *saw*. "Tot modis insitam arborem vidimus, omni genere pomorum ornustum: alio ramo nucibus, alio baccis, aliunde vite, ficis, piris, punicis, malorumque generibus. Sed huic brevis fuit vita." *

As we must not doubt that Pliny saw the specimen to which he here so pointedly alludes, we cannot otherwise explain the fact, than by supposing him to have been imposed upon by a practice which it is said is still resorted to in Italy, for amusement or deceit. The French have termed it the "Greffe des Charlatans." It consists in cutting down a tree, as the orange, to

* Lib. xvii. ch. 17. sect. 26.

within a short distance of the ground; then hollowing out the stump and planting within it several young trees of different species and families. In a few years the whole grow up together so as completely to fill the cavity, and on a superficial observation appear to have become blended or grafted into a single stem. The deception is still more perfect if a few buds have been left upon the stump to keep this alive also.

(228.) *Kinds of Grafts.* — M. Thouin has described about a hundred different ways in which the process of grafting may be varied. These may however be referred to the three following general classes.

1. *By Approach.* — Two plants are placed near each other, and their boughs grafted together whilst they are still on the stems. When they have become completely united, one is then severed from its own stock and left to grow on that of the other.

2. *By Slips.* — A shoot is taken from one tree and placed on the extremity of a branch of another properly prepared to receive it. The branch is cleft and the graft inserted into the notch in various ways, which more peculiarly form the study of the gardener. This graft is made in the spring when the sap is rising.

3. *Budding.* — A piece of bark is removed from a tree at a place where there is a bud; and a piece of the same dimensions is taken from another tree also containing a bud and is then placed on the exposed alburnum of the former tree. The branch is tied tightly above the graft in order to force the rising sap into it. This graft is practised both in spring and autumn.

(229.) *Effects of Grafting.* — It does not appear that the graft produces any decided effect upon the stock, as we have already remarked (art. 224.); but in certain instances the reverse seems unquestionably to be the case. The influence is rather to be attributed to some difference in the mode of growth in the two subjects, than to any dissimilarity between the two saps of the stock and graft. Thus the lilac grafted on the ash becomes a tree, and the *Mespilus japonica* on the hawthorn is capable of sustaining a greater degree of cold than it otherwise could. In some cases the crop of fruit is increased, in others it is diminished; and some plants which are naturally climbers become more bushy, &c.

(230.) *Development.* — The process of development never appears to be entirely stationary in the living plant, not even during winter when the repose of vegetable life is the most marked; but a slight progression of the sap is still going on and a trifling enlargement of the buds is gradually taking place. As the spring advances the vital energies revive and vegetation seems to awaken; a sudden and rapid flow of the sap towards the extremities takes place, and the buds begin to develop with great rapidity. It is evident that the increased temperature of the atmosphere is a stimulating cause in producing these effects; and they may be partially accelerated or retarded by artificial means. If for instance a branch of any tree growing in the open air is introduced into a hothouse during the winter, the buds upon it swell and put forth leaves although the rest of the tree continues bare.

(231.) *Vernal Development.* — The different degrees of vigour with which buds burst forth in spring in different years, is probably regulated by the quantity of nutriment which has been prepared and laid up in the stem during the previous summer; so that a more rapid development will take place after a fine season than after a bad one. The extraordinary activity which vegetation evinces in the spring, appears to depend upon the great freshness of those parts by which the several processes of nutrition are then conducted. New fibres have been formed at the roots during the winter, and their absorbing powers now act with the fullest energy; the young leaves have their vessels and vesicles quite fresh, and unobstructed by the deposition of those earthy matters which are afterwards found in them when the exhalation of moisture from their surface has been going on for some time. If a branch of the vine, sycamore, and many other trees be cut off at this period, the sap often flows with sufficient rapidity to fill a bottle in a few hours. As the summer advances this action gradually diminishes; but in the autumn it is again partially renewed.

(232.) *Autumnal Development.* — The buds formed in the axils of the leaves of many plants have attained by autumn a sufficient size to attract the sap towards them, and then they undergo a partial development, which however is soon checked on the approach of winter. In a few cases, as in the Lombardy poplar, this autumnal development is sufficient to furnish the extremities of some branches with leaves which remain for some time after the older leaves have fallen. This always takes place in mulberry trees in those countries where they are stripped for the purpose of feeding silkworms. The buds then become the centres of attraction to the rising sap, and soon developing furnish the trees with fresh leaves which replace those that have been removed. Such a tree lives as it were two years in one, but is always proportionably stunted and injured in its growth.

(233.) *Nutrition of Cryptogamic Plants.* — The higher tribes of cryptogamic plants possess true roots and leaves; and we may suppose their function of nutrition to be carried on in a way which differs little from that in which it proceeds among phanerogamic species. But the manner in which the lower tribes whose nutritive organs are not distinguishable into roots and leaves complete the function is in great obscurity, and few attempts have hitherto been made to elucidate the subject.

(234.) *Parasitic Plants.* — There are certain plants which are without the means of providing nutriment for themselves or of elaborating the crude sap into proper juice but obtain their nourishment immediately from other plants to which they attach themselves, and whose juices they absorb. Such plants are true "Parasites." They are distinguished from "Epiphytes," which also grow on the stems and branches of trees, but do not penetrate their bark or absorb their juices. There are a vast number of cryptogamic plants among the ferns, mosses, and lichens, which are epiphytic, as are also several species of certain phanerogamous tribes. This is particularly the case with those Orchideæ which are termed "air plants," whose roots imbibe moisture from the atmosphere as we noticed in art. 39. Among the true parasites, some cryptogamic species live wholly *within* the plant and may be considered analogous to intestinal worms; whilst such as are external (both cryptogamic and phanerogamic) may be likened to the ticks and lice which infest animals. Different species are parasitic on different parts of plants as on the root, stem, or leaves. Some of the cryptogamic species are highly destructive to our crops, as those which cause the "smut" and "rust" in corn. It is difficult to ascertain in what manner the impalpable powder into which their sporules disperse is introduced within the very substance of the plants attacked; but it seems not improbable that it may be imbibed with water by the roots. Some suppose it may be introduced through the stomata, but this is not so plausible an opinion as the former. All the phanerogamic species except those of the natural order Lorantheæ (to which the common misseltoe belongs) are destitute of green leaves; these organs appearing only in the form of small brown scales without stomata, and incapable of performing the functions of respiration. Hence these plants have a livid and discoloured appearance. They are furnished with suckers which penetrate the bark and absorb the proper juices of the plants on which they grow, and which are always dicotyledonous. It is remarkable, that the flower of largest dimensions hitherto discovered is a parasite of this description. This is the *Rafflesia Arnoldi* (*fig.* 159.) whose corolla measures a yard in diameter and is fifteen pounds in weight. It grows in the island of Sumatra upon the woody stems and roots of a trailing plant (*Cissus angustifolia*). In our own country the genera Orobanche, Cuscuta, Lathræa, Mono-

159

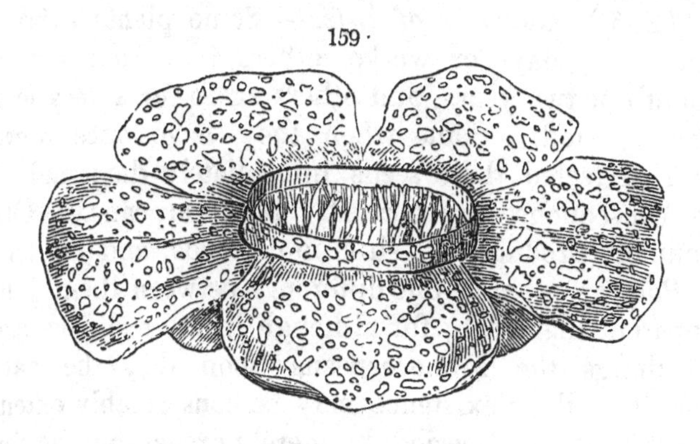

tropa, and Epipactis afford us leafless parasitic species. These do not appear to be very injurious to any woody plants which they attack; but such as grow on herbaceous species are highly mischievous. The species of "Cuscuta" are among the most curious of this kind. When they first germinate they have a stem formed like a delicate thread, which is leafless and soon coils itself round the stem of some plant growing in the neighbourhood. To this it adheres by means of suckers formed of wart-like protuberances at intervals along its stem. When it has obtained firm hold of the plant round which it has coiled, its root decays and the stem ceases to have any connection with the soil, but vegetates and produces flowers at the expense of the proper juices of the plant to which it is attached. The common misseltoe and other Lorantheæ being furnished with green leaves are able to elaborate crude sap into proper juice; but as they are destitute of any true root they possess the property of penetrating through the bark of the trees to which they are attached, and of fixing the base of their stems into the wood beneath. Thus they absorb the rising sap in its progress towards the leaf. It is asserted that a branch of misseltoe when placed in water has no power of absorbing this fluid, but that when the branch to which it is still attached is immersed, then the water is readily absorbed and penetrates into the misseltoe itself.

(235.) *Duration of Life.* — Some plants exist only for a few days or weeks, others for about a twelvemonth or two years, and others again for a very lengthened period. Some when they have once flowered and perfected their seeds immediately die; and these in consequence are termed "Monocarpeans." Others annually produce a fresh crop of seeds, and are termed "Polycarpeans." The difference between them is more apparent than real; for although in the ordinary course of things the Monocarpeans soon die, the natural period of their existence may be considerably extended beyond the usual period, by merely preventing the formation or development of their seed. This shows us that it was the effort of the plant to form seed which checked the functions of nutrition, and not that the period of its existence was necessarily so limited as its early death would seem to indicate. Some plants which are annuals in our stoves are perennials in their native country. The American aloe (*Agave americana*) is a striking example of a plant, the ordinary period of whose existence may be very considerably extended by preventing its flowers from developing. In its native climate it comes into blossom when four or five years old, and afterwards dies; but in our greenhouses it continues to vegetate for fifty or a hundred years without showing any symptoms of putting forth its flowers. If then we make abstraction of those checks which are given to the vital function by the process of fructification, and which do not appear formidable in any degree to the life of perennial species, we might imagine it possible for plants to continue vegetating for a much longer period than they naturally would; and that the life of some might be extended indefinitely, provided the external or accidental causes which tend to produce decay and death were continually removed. By this we mean, that certain plants never die from the effects of old age in the same sense in which we apply this term to animals, but are as well qualified to perform all their functions with vigour and precision after they have existed for many years as when they were young. The causes why such plants perish are not merely those common accidents which result from the influence of the weather, the ravages of animals, and the like external accidents, but likewise the continually increasing difficulty they meet with in procuring sufficient nutriment. The increasing length of their branches affords greater hold to the wind, and renders them proportionably more liable to be broken off and rottenness to be introduced in consequence. But in speaking of the duration of life in plants, we ought to have some definite notion of what we mean by a vegetable individual.

(236.) *Individuality of elementary Organs.* — Some persons consider every vesicle and other elementary organ of which plants are composed, to possess a distinct and separate existence of its own; and therefore they look upon every specimen as an aggregate of vegetable individuals, closely packed together and constituting a compound individual. The main facts upon which this singular hypothesis reposes are the following. — There are certain plants among the lowest tribes which consist of only one or at most of very few distinct vesicles, which indicates the possibility of a single detached vesicle existing as a separate individual. It may be observed however that these plants are among some of the most minute objects of organised matter, and that we know very little of their actual history and scarcely any thing of their physiology. Another argument in favour of the individuality of each vesicle is deduced from a belief that the cellular tissue in every part of the vegetable structure is capable of producing buds or gems, each of which is able to exist separate from the plant on which it was developed, and by proper treatment to become an individual plant similar to its parent. M. Turpin has recorded a very interesting and remarkable instance of this description, where a leaf of an Ornithogalum after it had been placed between some sheets of paper for the purpose of being dried for the herbarium, threw out a multitude of minute bulbs from all parts of its surface. He concludes that each separate bulb was only a more developed state of a single cell, and hence he would draw the inference that each cell must be a distinct individual. But if this conclusion were admitted, the same thing might be asserted of every organ which produces an embryo of any kind. It would perhaps have been more logical to have considered each cell as an embryonic sac, capable of originating a distinct individual of the same complicated form and structure of which it was itself only a subordinate organ. If each vesicle were an individual plant, its offspring if we argue from analogy ought to resemble itself, and to be a vesicle and not a bud with a complicated arrangement of parts representing in miniature the several organs of the entire plant. This hypothesis of the individuality of each vesicle according to our acceptance of the term appears to be untenable.

(237.) *Individuality of Buds.* — A second hypothesis considers each bud as a separate individual, possessed of a vitality independent of that of the whole plant. This view is considerably supported by the great analogy which exists between the structure of a plant considered in this light and that of some of the lower tribes of animals. The reproduction of polypi is effected by means of little bud-like protuberances on their surface, which having attained a certain degree of development quit the body of the parent and become separate individuals. Thus also if the buds on the stem of a tree are removed and treated with proper precaution, they will grow and become trees themselves. Some buds are detached by a natural process, and the plant is ordinarily propagated by this means. Thus the death and decay of the orange lily (*Lilium bulbiferum*) causes the little bulbs which are produced in the axils of its leaves to detach from the stem; and these upon falling to the ground become so many individual plants. The runners of the strawberry, decay when the buds at their extremities have obtained a firm root in the ground, and thus the parent plant becomes separated from the numerous progeny scattered around it. But the closest analogy between a plant, considered as an aggregate of individuals, and any living animal, is that which exists in certain marine tribes still lower in the scale of organisation than the polypi to which we have referred. A number of these animals are grafted and blended together into a compound mass, in which each still possesses its separate individuality, and is capable of existing in a detached form. It is by the joint labours of these compound animals that a coral reef is raised from the bottom of deep seas to the surface. The innermost and oldest parts of the reef consist of the untenanted cells of those animals which have died, whilst a fresh crop is continually developing towards the surface. Thus also in a tree, the oldest parts of the trunk and branches is composed of matter in a dead or dying state, and it is the newly developed portions alone which contain the living materials capable of performing the functions of vegetation. As these latter portions originate from successive crops of fresh buds, the analogy alluded to is very complete.

It has been further observed, that if each bud be not a separate individuality, we might, by grafting several buds on the same stock, produce a tree composed of a multitude of species; which would be an absurdity.

(238.) *Individuality of Plants.*—Any cutting, layer, or bud, which has been detached from a plant, and grown in an isolated state, always retains the exact peculiarities of the individual plant from which it was obtained; but a seedling, raised from the same plant, will frequently deviate more or less from the original type, and present us with certain peculiarities of its own. This fact appears to favour another hypothesis, distinct from the two already explained, which considers the vegetable individual, in the most usual acceptation of the term, as an entire plant which has originated from the development of a single seed. But this definition of an individual involves the seeming absurdity, that an organised being may consist of several detached portions, each of which may exist apart from the others. Thus a cutting from a tree is a part of the individual from whence it was taken; and though it may also become a tree, it is no more than the developed state of a portion of the former. Since all the weeping willows in Europe, for instance, are said to have originated from cuttings taken from a single tree; according to this hypothesis, there is no more than one weeping willow in Europe, and that also can only be a portion of one which may be still growing in Asia. But whatever be the speculations of physiologists, we must admit the truth of the remark, "that in ordinary parlance we require some more precise mode of expressing ourselves, when we would speak of the individual weeping willow which shades the grave of Napoleon at St. Helena, as being the same plant which decorates the tomb of J. J. Rousseau at Ermenonville, although each may probably have originated from the same embryo." But if we cannot, in the present state of knowledge, exactly determine the requisites which constitute the individuality of vegetables, and may possibly consider as a separate existence what in reality constitutes the duration of a succession of individuals; yet whilst we choose to put such a limitation to our ideas, we may speak of the duration of life in a plant as the real existence of an individual, whether this plant may have originated from a seed, bud, cutting, or from any other mode by which it could be propagated.

(239.) *Longevity of Trees.*—When we consider each separate plant as an individual being, there is this manifest and important distinction between the mode in which its life is maintained, and that in which it is continued in any animal; — the plant annually renews all the different organs by which its various functions are carried on, and which are consequently as vigorously performed in the oldest tree as in the youngest. But although the organs which every animal possesses are continually sustaining a certain degree of repair, yet they are gradually wearing out, or ultimately become choked up in old age; and thus a definite period is naturally allotted to the existence of the individual from this cause alone. But the period of life to which plants attain is no way dependent on these conditions; but is regulated by a combination of external causes and internal influences of a very different kind. Those trees are most likely to endure the longest, which grow the slowest, and which attain the least height in proportion to the diameter of their trunks; and the antiquity of some trees of this description appears to be prodigiously great.

(240.) *Estimation of the Age of Trees.*—It is only the ages of Dicotyledons which can be ascertained with any degree of certainty. In Monocotyledons the diameter of the tree is not enlarged by annual additions of fresh cylinders of wood, as is the case with the former, whose ages may be accurately ascertained by inspecting a transverse section of their trunks. By placing a strip of paper upon this section from the centre to the circumference, and marking it along the edge where it intersects the concentric circles on the section, a convenient register may be obtained, not only of the ages of different trees, but of their comparative rates of increase at different periods of their growth. As the pith is seldom exactly in the centre of the tree, the best mode of obtaining the average annual growth is by measuring the circumference of the trunk, and then calculating for the mean thickness of each layer by dividing the semidiameter by the whole number of layers. These measurements should be made at a little distance above the soil, generally about four feet, where the trunk is free from protuberances and of an average thickness.

Where a complete section cannot be obtained, a lateral incision may be made, and by counting the number of rings in the portion exposed, an approximation may be made to the whole number; care being taken to make allowance for the more rapid increase of the trunk in the early stages of its growth.

In other cases, some judgment may be formed of the ages of very old trees, by ascertaining the rate at which others of the same species have increased within known intervals of time, and by then applying the rule thus obtained to the tree in question. The observer must be cautious when he is examining very large trees, lest he should be deceived by several trunks having become blended into one.

(241.) *Examples of Longevity in Trees.* — As examples of the mode in which approximations have been made towards the ages of very old trees, we may mention certain individuals of the lime, yew, and baobab.

1. *The Lime.* — A tree of this description was planted at Fribourg in Switzerland, on the day when the news of the victory of Morat arrived, in 1476. In 1831, this tree was 13 feet 9 inches in circumference, which gives $1\frac{3}{4}$ lines in diameter per annum as the mean rate of its increase. But as this tree is confined in a town, we may allow 2 lines per annum as the rate of increase for other trees more freely exposed, whose ages we may wish to ascertain. Now, there is a lime near Neustadt on the Kocher, in the kingdom of Wurtemberg, which was of large dimensions in the year 1229; since it is stated in ancient records, that the city was rebuilt after its destruction in that year, "near the great tree." A poem, bearing the date of 1408, describes this tree as having its branches at that time supported by 67 columns. Evelyn, in 1664, mentions the number of columns then to have been 82; and in 1831 they had increased to 106. At this period, the trunk was 37 feet 6 inches and 3 lines (Wurtemberg measure) in circumference, between 5 and 6 feet from the ground. This, upon an

estimate of 2 lines per annum for its growth, would make it to be between 700 and 800 years old. But as it is certain that it has not increased for some centuries at so rapid a rate, it may fairly be considered as above 1000 years old.

2. *The Yew.* — M. De Candolle ascertained, by inspecting three yews which had been felled, that they had grown at the rate of 1 line in diameter per annum during 150 years; and that one of them had increased somewhat less rapidly during the succeeding century. The rate thus obtained, he applies to the growth of some English and Scotch yews, whose dimensions were given by Evelyn in 1666, and Pennant in 1770. Among these, is a yew which the former describes as growing in the churchyard of Braburn in Kent, which was 58 feet 9 inches in circumference, or 2820 lines in diameter; indicating by the above rule, as many years for its age. If now living, this tree, according to such an estimate, would be more than 3000 years old. It may be doubted from the following account, whether the rate at which the yew increases in England is not more rapid than in France. There are two fine healthy trees of this kind in the churchyard at Basildon in Berkshire, which, according to the parish register, were planted in 1726. In 1834 they were very nearly of the same dimensions, and the largest measured 9 feet 3 inches in circumference at 4 feet from the ground: this gives 444 lines for its diameter, or 4 lines per annum as the mean rate of increase for a century. It appears however by some other entries in the same register, that the tree had grown more rapidly during the former half of this period than it has done latterly. Taking these data as a guide for estimating the ages of some old yew trees in the churchyards of two neighbouring parishes, it would seem that De Candolle's calculations should be reduced by about one third, in order to obtain a more correct approximation than that which he has given for trees of this description. It was found,

for instance, that the layers of wood at different depths, in a hollow yew tree at Cholsey, Berkshire, varied considerably in thickness; and that some of those which had been very recently deposited were 2½ lines, whilst others, which were more than a century older, were only half a line in thickness. This tree is between 14 and 15 feet in circumference; and there is another in the churchyard of the neighbouring parish of Aldworth, which is more than 19 feet in circumference, which, estimated by De Candolle's rule, ought to be above 900 years old; but may rather be considered as nearer 600 years.

3. *The Baobab* (*Adansonia digitata.*) — The last example which we shall select, is that of the enormous baobabs, or monkey-bread trees of Senegal, whose great ages Adanson has attempted to estimate from the following data.

Thevet mentions, in his "Voyages aux Isles Antarctikes," in 1555, some "beaux arbres," which Adanson found to be 6 feet in diameter in 1749. He judged, from Thevet's expression, that these trees could not have been less than 4 feet in diameter at the time when he saw them; and this opinion was strengthened by observing the extent to which the letters of certain inscriptions upon them had become deformed, and which inscriptions were dated from the fourteenth and fifteenth centuries. Allowing therefore that these trees had increased 2 feet in diameter during two centuries, he estimated their age at 600 years. But there are trees of this species which are 30 feet in diameter; and these, at the above rate, would be 3000 years old. But if the age of these trees be calculated upon mathematical principles, it should seem that they must be much older even than this. Thus, Adanson having ascertained that a tree of 1 year old was 5 feet in height and 1 inch in diameter, and a tree of 30 years was 22 feet high and 2 feet in diameter, he applied these data to construct a table, which should give the heights and diameters of trees from 1 year to 5000 years old. From this we shall make the following extract: —

Age.	Height.	Diameter.
1 year.	5 feet.	$\frac{1}{12}$ feet.
30	22	2
100	29	4
210	40	6
660	53	11
1050	$58\frac{3}{4}$	14
2800	67	20
5150	73	30

It will be observed, according to this table, that the ages of trees whose diameters are 6 feet would be no more than 210 years; whereas it was satisfactorily shown that those which Thevet had described must at least be 600. So far then this table would underrate rather than exaggerate, the ages of these trees. It must be confessed that the estimate given for those of the largest dimensions is too startling to be received with implicit confidence; and that we need further evidence to satisfy us that these calculations are good approximations to the truth. Be this as it may, it seems to be sufficiently proved that the world is possessed of *living* monuments of antiquity, whose ages surpass those of the most stupendous fabrics which the labour of man has reared to perpetuate the memory of his folly or his superstition.

(242.) *Tables of Longevity of certain Trees.* — From various sources of information — some the results of direct observation, others the approximate values obtained from the kind of inferences which we have referred to — De Candolle has furnished us with the following list of remarkable trees, whose ages he considers that he has succeeded in ascertaining with some degree of precision: —

		Years.
1.	Elm - -	335.
2.	Cypress - -	350 (about).
3.	Cheirostemon -	400 (about).
4.	Ivy - -	450.
5.	Larch - -	576.
6.	Orange - -	630.
7.	Olive - -	700 (about).
8.	Oriental plane -	720 (and upwards).
9.	Cedar - -	800 (about).
10.	Lime - -	1076—1147.
11.	Oak - - -	810—1080—1500.
12.	Yew - -	1214—1458—2588—2820.
13.	Baobab - -	5150 (in 1757).
14.	Taxodium -	4000 to 6000 (about).

CHAP. V.

FUNCTION OF REPRODUCTION. — *Periods* 1, 2, 3.

PROPAGATION (243.). — ORIGIN OF FLOWER-BUDS (245.). — FLOWERING (246.). — FUNCTIONS OF THE PERIANTH (252.). — DEVELOPMENT OF CALORIC (254.). — FERTILISATION (255.). — FORMATION OF POLLEN (261.). — MATURATION (265.). — FLAVOUR AND COLOUR OF FRUIT (273.).

(243.) *Propagation.* — THERE are two distinct modes, according to which the propagation of the vegetable species is naturally secured, viz. "subdivision" and "reproduction." In the first the individual plant may be subdivided into several parts, each of which when detached from the parent stock is capable of existing as a separate individual. A familiar example of this mode of propagation may be seen in the common strawberry, to which we have alluded in art. 237. It is very common to find elms, poplars, and other trees throwing up suckers from their roots at a distance from the trunk, all of which are capable of becoming so many distinct trees, under favourable circumstances. Man has availed himself of this property, to extend the means which nature has provided for the propagation of the species; and by placing cuttings, slips, and buds under proper treatment, he forces them to throw out roots; or he grafts them on other stems, where they adhere and develop as so many separate and independent individuals. The process by which any detached portion of a plant becomes a distinct individual, similar to that from which it was derived, depends upon the power it possesses of reproducing those organs or parts in which it may be defective. Thus the ascending organs develop roots; and these again, produce buds from which the ascending organs proceed.

(244.) *Reproduction.* — But although the propagation of many plants may be effected by the means here alluded to, and although some species are more frequently and readily propagated by subdivision, than by the method which we are about to describe, yet the greater number of plants, and *at least* all those which bear flowers, secure the continuation of their species by a distinct process, of a very different nature. This constitutes the function of "reproduction," properly so called; which consists in the formation of seeds, containing the germs of future individuals. This function of reproduction is to the species, what life is to the individual — a provision made for its continued duration on the earth. The more minute details of the process by which the function of reproduction is carried on, and the germ or "embryo" of the future plant becomes generated in the seed, were never understood till of late years; nor are they even yet so completely ascertained as we may one day hope to find them. The general function of reproduction may be considered as completed in five different periods; much in the same manner as we ascribed seven periods or processes to the function of nutrition.

(245.) *Origin of Flower-buds.*—We find some buds capable of developing into branches and leaves, and others destined to produce flowers: but it is beyond the limits of our present faculties to ascertain by what law they are thus specially inclined, in their nascent state, to assume the one rather than the other of these characters. That leaf-buds and flower-buds have fundamentally the same origin, is apparent from an extensive review of those singular deviations from the ordinary productions of nature, which are termed Monstrosities, as we have already stated in art. 85. The organs developed from a flower-bud serve a temporary purpose, of a very different description from that assigned to those which are developed from a leaf-bud; and when that purpose is completed, they soon decay. The causes which predispose the plant to produce a flower-bud rather than a leaf-bud must begin to operate long before we are able to detect any traces of the bud itself; and from the very earliest period that we can perceive its existence, it has already assumed the peculiar characters with which it is destined to develop. It is asserted that in some palms, the flower-buds which are to produce flowers during seven successive years may all be detected at one time in the inner parts of the stem. We may further notice the manner in which the Lemnæ (*Duckweeds*) are propagated, as affording a striking argument in favour of the common origin of all buds. Each plant is a little green lenticular and frond-like mass, which produces a long pendent root from its under surface (*fig.* 31.). Its usual mode of propagation is by a bud or gem, which makes its appearance on the edge of the frond, and when fully developed, detaches itself and becomes a separate individual. In some seasons however, and under circumstances suitable to such an event, these plants put forth diandrous flowers, which originate precisely in those spots where the gems are usually developed.

FIRST PERIOD OF REPRODUCTION.

(246.) *Flowering.* — When the flower-bud is distinguishable, the parts of which the flower is composed are in a very rudimentary state. The perianth especially, continues for some time very small in proportion to the anthers, which are more early developed. A gradual enlargement of all the parts of the flower continues to take place till the period of expansion arrives. This expansion may be likened to the age of puberty in animals; and when completed, terminates the first period of the function of reproduction. In herbaceous plants, it is very frequently effected the same year in which they have germinated from the seed; but there are some which do not flower until the second year, and others not until later. Some undershrubs also begin to flower within the year; others not until after a second, third, or fourth has elapsed. Shrubs and trees, with very few exceptions, never flower before the second or third year at least, and very many of them attain a considerable age before they show any symptom of flowering. It may be asserted of trees, almost as a general rule, that the period when they commence flowering is protracted in proportion to the slowness of their growth.

(247.) *Stimulants to Inflorescence.* — Although we cannot comprehend the primary causes upon which the formation of the flower-bud depends, we can connect several phenomena which attend its development with the operation of specific influences. For instance, an increase of temperature accelerates, and a diminution retards the period of flowering; and according to the nature of the individual, these causes also operate in predisposing its buds to assume the character of leaf-buds or flower-buds. Many plants, when removed from a warm climate to a cold one, or *vice versâ,* will flourish without ever producing flowers; and others which are able to flower, never perfect their fruit. A superabundance of moisture retards the flowering, and also affects the formation of flower-buds; and it is generally observable, that where the functions of nutrition are forced into a state of unnatural excitement, the plant has an increased tendency to produce leaf-buds rather than flower-buds. Hence it is remarked, that when the fruit trees of temperate climates are transplanted to the warm and moist regions of the tropics, they frequently become barren, although they continue to push their shoots with vigour. To counteract this effect, a practice is resorted to in the East Indies, of laying bare some part of the roots, which checks the growth, causes the leaves to fall, and thus predisposes the plant to form flower-buds instead of leaf-buds. At the period of flowering, the vital energies of the plant seem to be called into extraordinary activity, and the organs of inflorescence are developed with great rapidity. An *Agave fœtida* which had vegetated in the Paris garden for nearly a century, and during that period had scarcely shown any signs of increase, during a warm summer began to show signs of flowering. In eighty-seven days, it had grown twenty-two feet and a half, and during one portion of this interval it increased at the rate of nearly one foot per diem.

(248.) *Periods of Flowering.* — The precise periods at which a species commences flowering in different years, range within certain limits, dependent partly upon the state of the weather; but it is very difficult to appreciate all the causes which concur in modifying them. It is evident that the annual distribution of temperature produces a marked effect upon the period of flowering, and that this operates more decidedly on those plants which flower in the spring, than on such as flower later in the year. The almond, flowers at Smyrna in the early part of February, in Germany about the beginning of April, and in Christiania not until the beginning of June. The vintage, however, takes place at Smyrna the beginning of September, and in Germany about the middle of October; a retardation in this case which is less than in the former.

When a perennial has once begun to flower, it is subject to periodic returns of this function. The period of the year in which the flower expands, is regulated in all cases by the peculiar character of each individual, and it is very nearly the same for all plants of the same species. There are, however, remarkable exceptions to the laws by which the periods of flowering in different species are regulated. Advantage is taken of this circumstance; and by propagating from such individuals as are both the earliest and latest in producing their seeds, peculiar "races" are gradually established, which secure to the cultivator a longer succession of a given crop than he could otherwise have obtained. De Candolle mentions an instance of a horse-chestnut at Geneva, which always flowers a whole month before the rest in its neighbourhood, without any apparent cause for such precocity. These anomalies indicate some peculiarity of constitution, or "idiosyncrasy," in the separate individuals; but they determine nothing against the existence of a general law, by which each species is supposed to be regulated in producing its flowers at a certain period of the year. A very abundant crop of fruit generally absorbs so much of the nutriment prepared in the stem, as to diminish, and often entirely to prevent the formation of flowers in the following season; and hence, some trees in orchards bear abundantly only on alternate years. As double flowers produce no fruit, their stems are not so thoroughly exhausted; and perennials of this description generally flower earlier in the season than single flowers of the same species. By far the greater number of plants flower in the spring, and several do so even before they expand their leaves. In these cases, the nutriment which has been prepared for the development of the flower, must have been wholly provided by the leaves of the preceding season, and have been magazined through the winter in the stem.

The peach, apple, and almond are familiar examples. It sometimes happens, when the leaves have been destroyed by drought or other causes, that a second crop of flower-buds is developed late in the year; the trees having sustained a check in their vegetation, similar to what takes place in the winter, break out again as if it were a second spring.

(249.) *Periodic Influences.*— The periods at which the flowering of plants commences in different years, at a given spot, appear to depend upon the mean distribution of temperature per month, rather than upon the mean annual temperature. Since some process or other of the function of nutrition is carried on throughout the year, and even in winter this is not entirely dormant, there may very likely be a critical season, when some defect of moisture, light, or temperature would be fatal to the progress and perfection of a particular process, and retard or completely prevent the flowering of the plant at the proper time. When by a combination of circumstances — partly dependent on the peculiar constitution of the individual, partly on the character of the species, and partly on external influences — the periodic return of a plant's flowering has been fixed within certain limits, to a given month in the year, it requires a certain lapse of time before any alteration in the external circumstances to which it may be subjected, can effect a decided change in this period. Thus, it is observed that plants which are transported from the southern to the northern hemisphere, do not immediately accommodate themselves to the opposite condition of the seasons in which they are placed, but for a while continue to show symptoms of flowering, at the same period of the year in which they had been accustomed so to do in their native climate. In some instances they are several years in accomplishing the change, and sometimes even die before they can effect it. The usual limits within which the periodic returns of flowering in each species take place, are always mentioned in the Floras of a given district; and

Linnæus and others have prepared tables of different plants, which flower in each month of the year, under the title of Flora's Calendars.

(250.) *Horary Expansion.* — As the flowering of different species takes place at different seasons of the year, so also many species open their flowers only at certain hours of the day. The greater number are not subject to any very marked law in this particular; and their flowers, when once expanded, continue open until they decay. Some flowers, as those of the purple horned-poppy (*Rœmeria violacea*), expand early in the morning, and their petals are so very fugacious, that they are mostly fallen two or three hours before noon. But there are many plants, as the *Convolvulus nil,* which retain their corolla for several days, and regularly open and shut it at certain hours. Linnæus prepared tables to express these facts, which he fancifully termed Flora's clocks. The following list may serve as a specimen.

A. M.

4. Convolvulus nil.
5. Papaver nudicaule.
5—6. Convolvulus tricolor.
6—7. Sonchus oleraceus.
8. Anagallis arvensis.
9. Calendula arvensis.
11. Ornithogalum umbellatum.
12. Mesymbrianthemum.

P. M.

2. Scilla pomeridiana.
5—6. Silene noctiflora.
6—7. Nyctago jalapa.
7—8. Cereus grandiflorus.
10. Convolvulus purpureus.

He named those flowers "Ephemeral," which open once only at a given time, and decay within the period of a day; and those "Equinoctial," which open and close for several days at the same hour. Of these, some are diurnal, others nocturnal. "Meteoric" flowers are such as are influenced by the state of the atmosphere. A few of these as the Calendula pluvialis close at the approach of rain; others as the Campanula glomerata when the sky is clouded.

(251.) *Stimulants to Expansion.*—Light and not heat appears to be the chief stimulus which regulates the expansion of the blossom; and the influences of moisture alone do not seem to affect it greatly; at least plants when wholly immersed in water expand as freely as in the open air. The phenomenon of their alternately expanding and closing, is allied to the sleep of the leaves (art. 155.), and like the periodic returns of flowering, appears to be regulated by the joint operation of several causes, among which we must allow that the peculiar idiosyncracy of each individual plays its part. For independently of the effect produced by the external stimulus of light, if a plant *accustomed* to flower at a given period of the day be removed to a dark room it will still make an effort to expand its flowers at the wonted hour. De Candolle proved this by shutting up some of these equinoctial plants, as Linnæus termed them, in a dark chamber by day and exposing them by night to strong lamp-light. This treatment occasioned for a while the greatest irregularity in their periods of expanding; but at length they became accustomed to the change, and closed their petals by day and opened them by night. In some cases the expansion of the flower is evidently influenced by the effects of light, heat, and moisture. The common dandelion (*Leontodon taraxacum*), when closed on a cloudy day, upon being brought into the stove will immediately expand its blossoms, though it may now be exposed to less light and more moisture than before. On the other hand, if the same plant be exposed to the light of the sun, it will also expand though the temperature may be lower than on a cloudy day, when it would continue shut. It has been often asserted and as frequently denied, that the common sunflower will continue to turn its blossoms to the sun during his diurnal course through the sky. That such is not always the fact is easily seen, for it often happens that a single plant is covered with blossoms, which face all quarters of the heavens. It is possible there may be some foundation for the opinion, and that under a more genial climate this may be the fact; or perhaps the notion may have originated in some confusion of ideas connected with the name of the plant, which seems at least as much entitled to its appellation from the appearance of its flowery disk surrounded by the glory of its golden rays, as from the very doubtful property which has been assigned to it. An effect of the kind alluded to is sometimes strikingly exhibited in such flowers as *Hypochœris radicata*, and *Apargia autumnalis;* which may often be seen in meadows where they abound, most evidently inclining their blossoms towards that quarter of the heavens in which the sun is shining.

(252.) *Functions of the Perianth.*—The universal presence of the stamens and pistils in every species of flowering plant, and the frequent want of a corolla and in some cases of a calyx also, appear to indicate that the functions of the two outermost whorls of the flower forming the perianth, are not so essential to the perfecting of the seed as the two innermost. In many cases indeed, where these whorls are not developed, some traces of their existence are nevertheless apparent in the form of glandular protuberances or nectaries; and it is possible that these may still perform whatever "function" more especially belongs to the perianth; just as the green surfaces of stems which do not develop leaves, perform the function of respiration. One obvious use of the calyx and corolla, when they are present, is to protect the inner whorls from injury in the early stages of their development. It seems not unlikely that they may primarily be destined in some way to modify the materials which are provided for the formation of the pollen and ovules. In addition to the purpose which the calyx and corolla serve, of protecting the stamens and pistils in the early stages of their development, they occasionally perform a similar office at a later period in protecting the seed. In some cases they remain attached to the seed-vessel in the modified form of membranous or chaffy appendages, which serve as sails to waft the seed to a distance. Some of the most familiar and effectual contrivances of this description are to be seen in the Compositæ; such as the common dandelion and thistles. In these cases the down attached to each seed is only a modified form of the calyx.

(253.) *Functions of the Nectary.*—As the nectary has been noticed in not fewer than seventy-two families, and is found in a vast number of species, its use is probably of some importance in the general economy of reproduction, though we do not know what this may be. The most plausible conjecture that has been offered supposes the secreted matter or nectar to be discharged by the organ on which it is seated or near which it is placed, whilst it is elaborating the juice for the use of the inner whorls. An important secondary purpose which it serves is to allure bees and other insects, which crawling over the flowers, and passing from one to the other, facilitate the dispersion of the pollen, and thus promote the fertility of the plant in the waywe are about to mention under our second period.

(254.) *Development of Caloric.*—At the time of the flower's expansion a considerable development of heat takes place in certain species, and there is also a rapid formation of carbonic acid. This phenomenon is most strikingly exhibited by some of the Arum tribe. The spadix of the common arum (*Arum maculatum*) attains a temperature of 7° R. or $47\frac{3}{4}$° Fahr. above that of the atmosphere, and the *Arum cordifolium* in the Mauritius has been observed to attain a temperature of 44° to 49° R. or 131° to $142\frac{1}{4}$° Fahr. that of the surrounding air being at 19° R. or $74\frac{3}{4}$° Fahr. These effects take place once only for each plant, and it seems most likely that they are the result of some chemical action, rather than of any physiological property.

SECOND PERIOD OF REPRODUCTION.

(255.) *Fertilization.*— Great progress has been made within the last few years towards attaining an accurate knowledge of the process by which the fertility of the seed is secured. It had been long ascertained, that the action of the pollen was somehow essential to this purpose, and that the effect was also produced through the intervention of the stigma; but the manner in which it took place was not understood. Even the ancients had obtained some vague notions on the subject, although their speculations regarding this as well as most other minute details in natural science were replete with error and absurdity. The general fact had forced itself upon their attention in the cultivation of the date-palm. As the blossoms of this tree are diœcious, the distinction between those individuals which continued barren and such as always bore fruit was of course soon remarked; and it was found to be necessary that either some of the barren kinds should be cultivated in the neighbourhood of those which bore fruit, or else that bunches of their flowers should be suspended near them, otherwise the fruit never attained perfection. Hence originated the custom of cultivating only fertile plants, and of annually bringing bunches of the sterile flowers from the wild trees — a practice which has prevailed from the earliest periods of history to the present day in Egypt, and those countries of the East where the date forms a most important article of human food. When the French were in Egypt in 1800, the events of the war prevented the inhabitants from procuring the blossoms of the sterile or male plant (as it is considered) from the deserts, and none of the cultivated plants in consequence bore any fruit.

(256.) *Erroneous Theory of the Ancients.*— A practice has long prevailed in certain countries of the East with respect to the cultivated fig, of a similar description to that which is employed to fertilize the date, and although the results are very different in the two cases, it is only lately that this fact has been suspected. Both phenomena were always considered of the same class; and an erroneous theory was formerly founded on the mistake. Bunches of the flowers of the wild fig are brought from the woods and suspended over the cultivated plants, when a small insect (the larva of a cynips) imported with the wild flowers punctures the young fruit of the cultivated individuals, and accelerates their ripening — in the same way that we find a similar effect produced in some apples and pears by the puncture of the caterpillar of a small moth, which causes them to ripen before the rest, and to fall sooner from the tree. In consequence of the earlier ripening of the figs occasioned by the practice alluded to, and which is styled the caprification of their fruit, a second crop is secured which might otherwise have failed, from being produced too late in the season to allow of its attaining perfection. It was in attempting to generalise from the facts observed in the caprification of the young fig, that the ancients asserted that a maggot (ψην) was the efficient cause of fertility in the date, and that this insect crept from the sterile into the fertile blossoms before the development of the fruit could take place.

The existence of a sexual distinction between individual trees in such species as the date and some other diœcious plants, gave rise to another erroneous opinion, and it was supposed that even plants where the stamens and pistils were contained in the same flower were nevertheless unisexual. Thus Claudian asserts —

> "Vivunt in venerem frondes, arborque vicissim
> Felix arbor amat; nutant ad mutua palmæ
> Fœdera populeo suspirat populus ictu;
> Et platani platanis, alnoque assibilat alnus."

(257.) *Vegetable Sexes.* — A more careful research and the results of direct experiment have superseded the vague conjectures of the old philosophers; and it is now clearly established that the two innermost floral whorls, the stamens and pistils, are the organs essential to the fertility of the seed. In the case of double flowers where all the stamens have assumed the condition of petals, seed is never produced ; but if the pistil be perfect, it may be supplied with pollen from another plant of the same species, and will then ripen its ovules. Some apparent anomalies are recorded among the various experiments which have been made to prove the necessity of the action of the pollen in securing the fertility of the seed. The females of certain diœcious plants have matured their seeds although they were carefully excluded from the action of the stameniferous individuals; but in some of these cases, this was probably owing to the fact that diœcious plants are frequently partially monœcious, and that a stameniferous flower is here and there developed on the fertile plants, which may have furnished sufficient pollen to set the fruit. According to some recent experiments, however, the universality of a law which establishes the necessity of the pollen's action has been rather shaken, unless there be some error which it is difficult to account for. If they are correct, it seems to have been proved that hemp and a few other *annual* diœcious species are capable of ripening their seed without the action of the pollen having taken place. Even if the fact should be satisfactorily established it will in no way disprove the general necessity of the pollen's action, or the sexual distinctions of all phanerogamous plants. But such isolated exceptions may possibly be considered analogous to the case of the Aphides, in which insects a single impregnation is sufficient to enable several generations to become fertile. But after all we have such marvellous accounts of the distance to which the pollen may be carried and yet preserve its proper influence, that it seems hardly possible to feel quite certain that the plants in question may not have been fertilized from others growing in the neighbourhood. It is stated that in the year 1505 there was a female date-palm growing at Brindes, which flowered regularly but never bore fruit. At length a male plant of the same species growing thirty miles off at Otranto, having attained a sufficient height to overtop the trees in its neighbourhood, its pollen was then wafted by the wind across the intervening space, and the tree at Brindes produced its fruit. The poet Pontanus who flourished at the time, has also recorded the fact. The late colonel Wilkes when governor of St. Helena, procured some pollen from dates growing on the continent of Africa, with which he fertilized some trees on the island that had never before perfected their fruit. It is certainly not necessary that the ripe pollen should immediately be brought into contact with the stigma; and instances are recorded of its having been sent in a letter from one part of the country to another and still retaining its activity. Dr. Graham mentions that a female specimen of the Chinese pitcher-plant (*Nepenthes distillatoria*) was fertilized in the Edinburgh Botanic Garden, by pollen thus procured from a male plant which happened fortunately to be in flower in another part of Scotland.

(258.) *Dispersion of Pollen.* — Before the pollen is scattered from the anther, some plants seem to make preparation for increasing the certainty of its taking effect, by bringing the stamens nearer to the pistil. This is remarkably evident in the Grass-of-Parnassus (*Parnassia palustris*), whose stamens on the first expansion of the flower are inclined away from the pistil, but are afterwards brought in succession towards it when their anthers are about to burst. In Geranium, Kalmia, &c. the filaments bend until the anther is placed immediately over the stigma. In the berberry (as we have described in art. 149. 3.), the filament may be caused to incline suddenly towards the stigma by gently touching it near the base on the inside. The genus Stylidium affords one of the most singular examples of this kind of floral irritability; though in this case the object is not so clearly to be perceived, since the anthers are at first close to the stigma, and the pistil is suddenly removed from them.

But independently of any means which some species employ for assisting the dispersion of the pollen and securing its contact with the stigma, we find that the mere conditions in which the flower is placed are often such as are most likely to secure these results without further contrivance. Thus, when the flower is erect and the stamens are longer than the pistil, the pollen on falling from the anthers is most likely to come in contact with the stigma placed immediately below them; so also where the flower is pendent and the stamens shorter than the pistil, the same effects will be produced. In cases where the flower is erect and the stigma stands higher than the anthers, there is often a closer aggregation of the flowers as in the numerous order Compositæ, so that the chances are greatly increased whereby the pollen from one flower may be brought into contact with the stigma of another, either by the action of insects crawling over them or by the mere agitation of the wind. These and a thousand other instances might be adduced of a provision made for securing the perfect success of an operation of so much consequence to the preservation of the species.

(259.) *Protection of Pollen.*— It is further essential that the pollen should be protected from the influence of moisture; and, consequently we find that aquatics, as the water-lily (*Nymphæa alba*), elongate their flower-stalks until the blossoms float upon the surface of the water. In the water-soldier (*Stratiotes aloides*), water-violet (*Hottonia palustris*), and others, the entire plants float to the surface of the water during the period of flowering, but live submerged at other times. In the *Zostera marina* the flowers are arranged within a cavity filled with air: and thus, although they are developed beneath the surface, they are protected from the immediate contact of the water. But of all instances that might be mentioned, where the action of the pollen is secured by some singularity of structure or contrivance, the

Valisneria spiralis is one of the most remarkable. This is an aquatic, native of the south of Europe. Its flowers are diœcious. The females are attached to long peduncles which at first are spirally twisted, so that the buds are completely submerged. They afterwards untwist until the buds reach the surface, and the flowers expand. The males on the other hand have very short peduncles, and their buds are in the form of little bladders which easily detach themselves from the peduncle and float to the surface of the water when the pollen is ripe. Here they surround the female blossoms and then expand. The peduncles of the female plants coil up again, the flowers are submerged and the seed is then ripened below the surface of the water.

(260.) *Certainty of Reproduction.*—No one who feels as he ought the lessons which the study of nature is calculated to convey, but must be struck with admiration at witnessing the multifarious resources, combined with an extreme simplicity in the means employed, for effecting that unity of purpose which is manifested in the preservation of the numerous species that clothe and beautify the surface of the earth. Independently of that security which every species possesses in its reproduction by seed against the probability of utter annihilation, some are further enabled to maintain their position by means of creeping stems. Many aquatics, as the potamogetons, are thus extensively propagated at the bottom of rivers and lakes and their perpetuity secured, even though the conditions necessary to enable them to perfect their seed should never be fulfilled. On the other hand the occasional production of seed in such plants seems to be necessary, if we remember that their native bed may possibly be drained in the lapse of ages by one of those events which characterise the geological history of our planet; when the only chance which would possess of being preserved must consist in the probability of some of those seeds which they had "cast upon the waters," finding a new station equally congenial to their growth. The chances which threaten the failure of seed in diœcious species are diminished by the occasional development of a few flowers of an opposite sex among those which otherwise characterize the separate individuals; and it is well authenticated that cases occasionally occur, where willows which for many years had borne flowers of one sex only, have afterwards changed their character and begun to bear only those of an opposite sex.

(261.) *Formation of Pollen.*—Before we describe the action of the pollen, we shall say a few words upon its formation. In this case, as in the whole account of the fertilization and development of the ovule, we are especially indebted to the admirable researches of Adolphe Brongniart, who in a memoir published in the "Annales des Sciences," has combined an extensive series of original observations with whatever was previously known on the subject; and placed the main facts of this interesting and curious question beyond the possibility of successful contradiction. To Robert Brown also in this as in every department of botany, we are preeminently indebted for important and accurate details. His invaluable papers on the fecundation of Asclepiadeæ and Orchideæ form an important epoch in the progress of *general* physiology.

So soon as the anther can be distinguished in the flower-bud, its cells are filled with a mass of cellular tissue, each vesicle of which contains one or more grains of pollen. As the anther ripens these grains enlarge and ultimately rupture the vesicles; and the débris of the cellular tissue then forms loose fibres intermixed with the pollen. In general the grains are separate, but in some plants (as the heaths) three or four grains always adhere together. There is no appearance of any thing like a pedicel to the separate grains, nor any scar upon them like the hilum on the ovule, which might indicate an original attachment to the sides of the vesicles within which they were formed. In most plants each grain is composed of two membranes; the exterior presenting the various appearances described art. 99.; and the interior being an exceedingly delicate homogeneous pellicle. Whatever may be the ultimate determination of botanists, respecting the formation and origin of pollen, yet as its grains in a very early stage of their development are free and unattached to the inner walls of the anther, it should seem that from this period at least their growth must depend upon the absorption of nutriment through their surfaces.

(262.) *Action of Water on Pollen.*—If ripe pollen be placed in a drop of water and examined under a microscope, in a few seconds it will be seen to dilate, burst, and violently expel a cloud of minute granules (*fig.* 160.). These granules are still contained within the inner membrane of the pollen grain protruded through the ruptured outer membrane, but which is difficult to be observed, on account of its extreme tenuity. It thus forms a sort of rude sack, termed a "pollen tube," and contains a liquid, the "fovilla," in which are dispersed a number of very minute "pollen granules." The outer skin of the grains is ruptured irregularly in most Monocotyledons; but in Dicotyledons there are one or more determinate points on its surface where a regular dehiscence takes place, and it is through these that the inner membrane then protrudes. In consequence of the effect thus produced on pollen by water, it is liable to injury in rainy seasons and the fertility of the seed is often impaired. Although the granules are destined to convey that influence to the ovule which is necessary to secure its fertility, yet their violent expulsion from the grains is not the manner in which this effect is produced. This process constitutes one of the most curious phenomena which have been observed of late years among the many wonders which the microscope has brought to light. Considering the minuteness of the objects and the delicacy of the manipulations requisite for these investigations, we must feel surprised

160

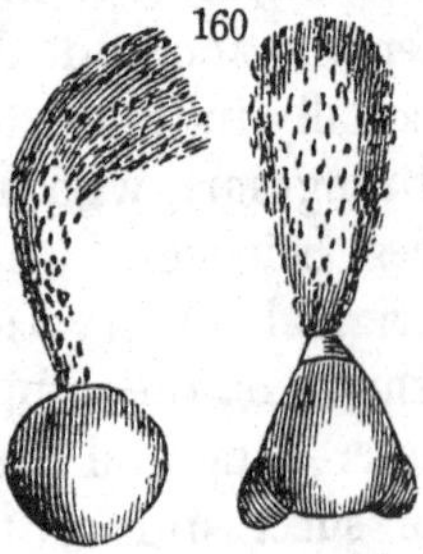

at the progress which this inquiry has already made, although much yet remains to be done before a complete elucidation of all points can take place.

(263.) *Granules.*—With a lens which magnifies about 300 times in linear measure, the form of the granules in the fovilla may be clearly distinguished. Whilst still in the pollen tubes they are often in motion, like the globules in the stems of the Chara (art. 194.). A few larger molecules are found dispersed among them, apparently of an oleaginous nature. In the same species all the granules are nearly of the same size and shape, but they differ in different species. They are always more or less spheroidal or cylindrical. They are certainly to be considered as the direct agents employed in securing the fertility of the ovules.

(264.) *Action of the Stigma.*— When the grains of pollen fall upon the stigma, they become attached to it by means of a glutinous exudation with which it is covered. No immediate action takes place, and the grains are not violently exploded with the pollenic tubes as when they are placed in water; but after they have remained for a few hours, and in some cases even for a few days on the stigma, each grain protrudes one or more delicate pollenic tubes which penetrate between the vesicles of the cellular tissue of the stigma (*fig.* 161. *a*). These tubes increase rapidly in length, growing as it should seem by means of the nourishment which they derive from the granular matter abounding in the interstices or intercellular passages between the vesicles of the style. In some cases if not in all, the pollen tubes become extended down the whole length

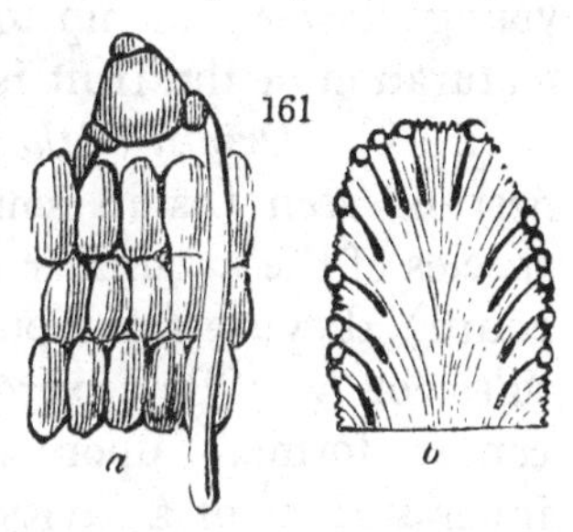

of the style, and penetrate into the cavity of the ovarium, where they run along the surface of the placenta, and surround the ovules. At (*b*) we have the section of a stigma on whose surface are numerous pollen grains each protruding a tube and appearing like pins on a cushion. In certain families, as the Orchideæ (*fig.* 162. *a*) and Asclepiadeæ (*b*), the grains contained in one cell of each anther are agglutinated together into waxy masses, so that when the action takes place, a number of tubes are protruded together and form a thickened cord (as at *c*); and thus they penetrate into the ovarium "en masse." Even some grains which are composed of only one vesicle, exsert more than one pollen tube. In some cases the tube originates in a swelling on the surface of the grain, which then seems to be formed of one skin only, or perhaps the two may be united.

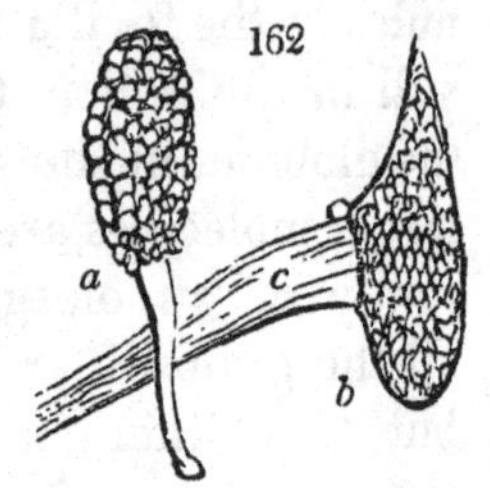

THIRD PERIOD OF REPRODUCTION.

(265.) *Maturation.*— After the action of the pollen has taken place, the ovules contained in the ovarium begin rapidly to increase, and the fruit swells and ripens. But in order to understand the several parts of which the seed is composed, it is necessary to trace the changes which the ovule undergoes, from the earliest period in which it is distinguishable in the young flower-bud, up to the time when the complete maturation of the fruit is effected.

(266.) *Origin of the Ovule.*— When the ovules can first be seen (as in some species of the cucumber or gourd), they are small pustules or wartlike excrescences formed upon the inner surface of a cavity in the ovarium; and are without any distinct traces of organisation (*fig.* 163. *a*). Soon after their first appearance we find them lengthening (*b*), and assuming traces of an organised structure (*c*). They are observed to consist of an internal mass of cel-

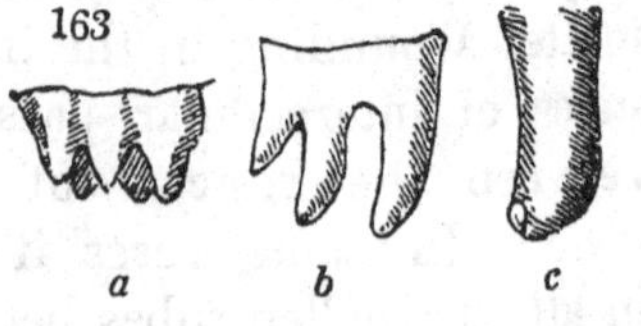

lular tissue termed the "nucleus" (*fig.* 164. *a*), invested by two coats or skins (*b*), open at their lower extremity, and allowing a portion of the nucleus, called its "apex," to protrude through them. This opening is termed the "foramen." Shortly afterwards these skins close over the nucleus, and leave only a small orifice to the foramen (*c*). The outermost of these skins is termed the "testa" or "primine," and the innermost the "tegmen" or "secundine." Sometimes there is only one skin, or more probably the two are so blended together that they are not distinguishable. As the ovule enlarges, the nucleus itself is also found to be a closed sack, of a thick or fleshy consistency; and within this and towards its apex, another small sack or vesicle makes its appearance called the "embryonic sack" (*fig.* 165. *a*). The ovule may therefore generally be considered in its early state to be composed of two *closed* sacks which together constitute the nucleus, and of two *open* sacks which form its integuments. In some cases the two outer skins appear to be blended together, for one only can be seen. The number of sacks which compose the nucleus sometimes also amounts to three; so that the whole number contained in the ovule is as many as five, and these have received the several names of primine, secundine, tercine, quartine, and quintine — reckoning from without, inwards. Whilst the enlargement of the ovule proceeds, a change of position also takes place in the relation of its parts, owing to an unequal development of the sides of the primine. The apex, which at first was on the side of

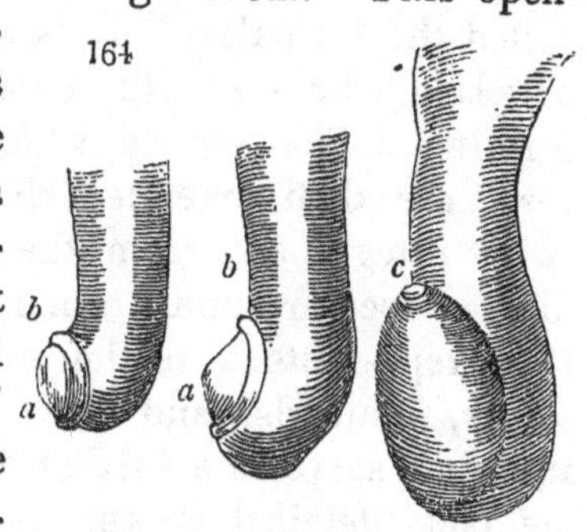

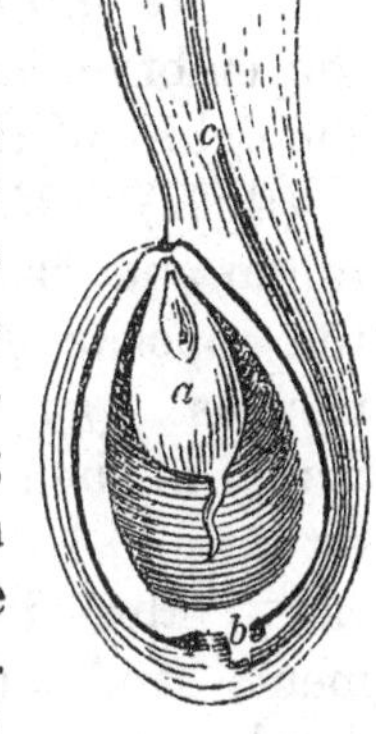

the ovule opposite the part by which it is attached to the ovarium, has now by some torsion of the mass been brought close to its base. In this case the point where the secundine is attached to the primine (and which is called the "Chalaze" *b*) is distinct from the "Hilum," or place where the funicular cord is attached to the primine. The vessels which penetrate the funicular cord, are then extended through the substance of the outer integument from the hilum to the chalaze and form a vascular bundle termed the "raphe" (*c*). Figure 166. represents a section of the developing ovules of plums, almonds, and other stone fruits, and may serve as a further illustration of the facts detailed in this article. When the embryo (*a*) makes its appearance in the embryonic sack (or quartine) (*b*), this latter organ is observed to be connected with three or four other large vesicles in communication with the raphe where it joins the chalaze (*c*); the hilum being at (*d*). The testa and tegmen already appear as one skin (*f*). The thick nucleus (*e*), together with the embryonic sack, are ultimately exhausted by the development of the embryo, and the spermoderm is then composed of the débris of the four integuments.

166

(267.) *Modifications of the Ovule.* — When the hilum and chalaze are contiguous and the foramen at the opposite extremity, the ovule is called "Orthotropous" (*fig.* 167. o), and this is the condition of all ovules in their earliest state. In many cases the integuments and nucleus develop more rapidly on one side than on the other, and a peculiar torsion takes place in the body of the seed, by which means the apex is brought near the hilum. The ovule is then termed "Campulitropous" (c). When the

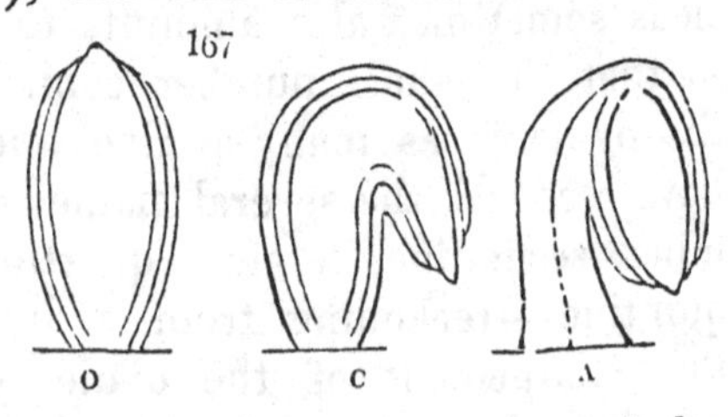

167

chalaze is removed from the hilum, so that the whole nucleus is inclined upon the axis, as described in art. 266. the ovule is termed "Anatropous" (A). It more frequently happens that the chalaze is immediately opposite to the hilum, and the foramen near it (as at A); but sometimes the former is placed on one side, at about a quarter of the circumference of the ovule.

(268.) *Formation of the Embryo.*— Such is the state of the ovules previous to the action of the pollen upon the stigma. Sooner or later after that action, the embryo makes its appearance under the form of a minute vesicle, attached to the summit of the innermost or embryonic sack, with the radicle directed towards the foramen, and the cotyledons towards the chalaze. It gradually enlarges, and the whole ovule also continues to increase.

(269.) *Formation of Albumen.*— Whilst the ovule is increasing, the testa and tegmen gradually part with their juices, for the support and increase as it should seem of the nucleus; and these two integuments are ultimately blended together, and their debris then forms only a single skin over the ripe seed. The nucleus itself is sometimes exhausted in a similar manner; whilst, in some cases, a deposition of nutritious matter takes place within the tercine, and round the quartine or embryonic sack. In some kinds of seed the nutriment thus provided for the embryo is secreted within the embryonic sack, and in others there is a secretion of this description going on simultaneously within this sack and the tercine also. In many cases this nutriment, or "amnios," as it is styled in its earlier state, is not wholly absorbed by the ripening ovule; and it ultimately becomes the "albumen" or "perisperm" of the seed, and is then farinaceous, hard, or oily. This superabundant supply of albumen is of further service to the embryo during its germination, and supplies it with nutriment in the early stages of its development, before the roots have sufficiently enlarged to absorb the sap from the surrounding soil. But in many cases there is no separate provision of albumen in a detached form, but this material, or something like it, is diffused through the substance of the cotyledons.

(270.) *Development of the Ovule.*— So soon as the embryo makes its appearance it becomes a centre of vital action, attracting the juices of the plant and beginning an independent existence. It continues to increase at the expense of its several envelopes, and in the end constitutes the bulk of the seed. The seed then consists of this body enveloped by a single skin (the spermoderm, art. 109.), which is composed of the débris of all the envelopes blended together, and in some cases there is also superadded a store of albumen. Those ovaries which are not fertilized soon wither up; but still it often happens that the ovaria containing them do not perish. On the contrary in some fruits,— as in the cultivated varieties of the pine-apple, where the ovules are universally abortive, — the ovary is developed into a fleshy pericarp; although such is not the case with the wild plants which possess ovules. The same is true also of the bread-fruit. In some oranges whose ovules happen to be abortive, the flavour of the fruit is much improved; but in many plants, when the ovules are abortive the ovary does not increase. In ovaria which contain numerous ovules it often happens that some only are fertilized; and sometimes only one ovule arrives at perfection, the rest being either starved for want of sufficient nutriment, or choked by the more rapid growth of that which becomes a perfect seed. In the oak for example, five ovules out of six are constantly abortive. In the horse-chestnut it seldom happens that more than one arrives at perfection, though the pericarp originally contained six; and though all of them, for some time after their fertilization was secured, had every appearance of health and vigour. In the stone fruits — plums, peaches, &c. — we generally find only one ripe kernel, though two ovules are always present in the early stages of the fruit; the other may be seen in a withered state attached to the inner edge of one suture of the stone, whilst the perfect seed is attached to the other.

(271.) *Maturation of the Fruit.*—Whilst the fruit continues to swell, the sap is drawn with increased energy towards those branches on which it hangs, and a rapid exhaustion takes place of the nutritious materials previously deposited in the stem. As these materials are distributed among the whole of the fruit, the advantage of thinning it early is evident, as the share which each will receive must be proportionably increased. We may compare the maturation of the fruit to the period of gestation in animals; and it is of very varied duration in different species. The greater number of plants ripen their fruit considerably within a year from the time when the flowers first expand, and some require only a few days for this purpose. But there are certain trees, as some oaks, which require eighteen months; and the fruit of the juniper, and the cones of many of the fir tribe, hang above a twelvemonth. The cedar requires twenty-seven months to bring its seed to perfection.

The following list contains a few other examples of the different periods required by some plants for the maturation of their seeds:—

Days	13.	Panicum viride.
	14.	Avena pratensis.
	16—30.	Most other Gramineæ.
Months	2.	Raspberry, Strawberry, Cherry, Elm, Poppy, &c.
	3.	Bird-cherry, Lime, Reseda-luteola.
	4.	Whitethorn, Horse-chestnut.
	5—6.	Vine, Pear, Apple, Walnut, Beech.
	7.	Olive.
	8—9.	Colchicum autumnale, Missletoe.
	10—11.	Most Fir trees.

No uncombined water is found in the seed when it is completely ripe; but it is now chemically united in their fecula, oils, &c., and the proportion of carbon also is then at a maximum. Hence it acquires an increased power of resisting decomposition, and of preserving its vitality under every temperature to which it is likely to be naturally exposed.

Most ripe seeds are of greater specific gravity than water, unless (as in the common Indian cress, *Tropæolum majus*) air happens to be contained in their envelopes, when they will float.

(272.) *Stimulants to Maturation.*—An increase of temperature materially accelerates the period in which fruits ripen, and also improves their flavour. Advantage is taken of this fact to wrap fruit in thin bags, to place it under glass, or upon slates of a dark colour. That elaboration of the juices by which the fruit is ripened is a local operation, and takes place within the fruit itself. This is clearly shown where a tree, whose fruit possesses a peculiar flavour, has been grafted upon the stock of another kind whose fruit possesses a very different quality: no alteration is produced upon the graft. Also where fruit has been gathered before it was quite ripe it will nevertheless ripen, as every one is aware is the case in apples, oranges, and many others.

The process of ringing the branches or stems of fruit trees, already alluded to in art. 190., considerably accelerates, as well as secures the maturation of the fruit. In the vineyards of France this has been practised on a large scale, and a peculiar instrument invented for the purpose; and the results have shown that the operation accelerates the ripening of the grapes from twelve to fifteen days. De Candolle mentions a vine near Geneva which regularly flowered every year, but had never produced fruit until this operation was performed upon it; and then the fruit set, and proved to be the small Corinth grape, which in commerce is known under the name of dried-currants or plums.

(273.) *Flavour of Fruit.*—We are wholly unacquainted with the physiological causes upon which the different flavours of fruits depend. In the earlier state of the pericarp, its functions are analogous to those of the leaf; but when this organ possesses no stomata and becomes succulent, at first there is a superabundance of water, but in ripening, an increase of saccharine matter takes place accompanied with a diminution of the water.

The percentage of water and sugar in the following fruits, in their unripe and ripe state, has been thus stated, viz.: —

	WATER.		SUGAR.	
	Unripe.	*Ripe.*	*Unripe.*	*Ripe.*
Apricot - -	89.39	74.87	6.64	16·48
Peach - - -	90.31	80 24	—	—
Red Currants -	—	—	0·52	6·24
Cherries (royales) -	—	—	1·12	18·12
Plums (reine-claude)	—	—	17·71	24·81

The solid portion of succulent fruits consists of lignine; and their liquid parts are chiefly water mixed with gum, malic-acid, malate of lime, colouring matter, and vegeto-animal matter. The whole is flavoured with an aromatic substance peculiar to each fruit. Much wet weather renders these fruits insipid; and many autumnal fruits acquire more flavour if they are detached from the tree before they are perfectly ripe.

(274.) *Colours of Fruit.*— The peculiar colours of fruit depend upon some local secretions, of which we are not able to give an account, any more than of those which produce the colour of the flower. These two phenomena have this property in common, that those parts which are usually coloured may become white in certain varieties, which may be propagated by slips and cuttings; even races of white-flowered and white-fruited varieties may to a certain extent be established by seed. The colours are deepened by the action of light.

CHAP. VI.

FUNCTION OF REPRODUCTION CONTINUED. — *Periods* 4, 5.

DISSEMINATION (275.). — MODES OF DISSEMINATION (279.) — PRESERVATION OF SEED (281.). — GERMINATION (283.). — VITALITY OF THE EMBRYO (290.). — RELATION OF BUD AND EMBRYO (291.). — PROLIFEROUS FLOWERS (292.). — HYBRIDS (295.).

FOURTH PERIOD OF REPRODUCTION.

(275.) *Dissemination.* — THE manner in which the ripe seed is disseminated, forms a more important element in the history of the preservation of species than might at first be imagined. It may be considered analogous to the period of labour in the animal kingdom, and still more strictly to the laying of eggs among such as are oviparous. If the different modes of dissemination were not in harmony with the peculiar character of the species, we might expect in the lapse of ages that some combination of circumstances would arise which should so far interfere with the reproduction of a given species that it would disappear from the earth. This is guarded against by some peculiar adaptation of the mode in which the seed is disseminated to the conditions under which each species naturally thrives the best. In some cases, the seed falls immediately around the parent plant; and where many seeds are contained in the same seed-vessel, the young plants come up in a crowded manner and occupy the soil in society, to the exclusion even of more robust species. Other seeds and seed-vessels are furnished with the means of being transported by the influence of the wind or by some other cause to a considerable distance. The great diversity in the means by which the dissemination of the seed is naturally secured forms one important inquiry to the botanical geographer; and a complete description of the various appendages by which their dispersion is assisted would form an interesting topic of inquiry. We may just refer to three forms of fruits which are more especially connected with the physiology of our subject, and which exercise a marked influence on the dissemination of the seed.

(276.) *In pseudospermic Fruits.*—In this class we may include all fruits whose pericarp is so closely attached to the seed, that it cannot readily be distinguished from one of its integuments. These are often erroneously considered as naked seeds, and not as complete fruits. To this class belong the various kinds of corn; the seeds of the umbelliferæ, as carrots, parsnips, &c.; and of the compositæ and others. In these cases, the seed is sown together with the seed cover (or pericarp), and the young plant has this additional obstacle to overcome before it can grow. Many fruits of this kind are furnished with wing-like appendages, as in the ash and sycamore; or with down, as in the valerian, but more especially in some of the compositæ, as the dandelion, thistles, and others. All these contrivances are manifestly intended to assist in the dissemination of the seed; but in many cases the pseudospermic seeds have no such provision, and are even so arranged on the plant as to secure it against any very extended dispersion.

(277.) *In fleshy Fruits.*—The soft pulp which surrounds the seeds of fleshy fruits does not appear to accelerate their growth when sown with them; and by its tendency to rot, it prevents them from keeping so long as when they are divested of it. As a sort of compensation for the injuries which they might receive on this account, many seeds of pulpy fruits are encased in a hard stone or bony envelope which resists the action of moisture, and protects them from the influence of the rotting pulpy mass on the exterior. All fruits of this kind fall to the ground close to the plant which bears them, and must depend upon accident for their dispersion; but as nature has destined these fruits to be the favourite food of many birds and other animals, they become instrumental in doing this. Animals after swallowing these fruits digest the pulp only, whilst the seed is voided by them in a state better fitted for germination than it was before.

(278.) *In capsular Fruits.*—Under this denomination may be included all fruits whose pericarp consists of a dry cover, which generally becomes detached from the seed, and bursts regularly along a line of suture, separating it into distinct valves. Most of these fruits are many-seeded, and their dispersion is commonly effected by the agitation of the wind, which shakes a few at a time from the capsule. In some cases they are so arranged that their dispersion is necessarily protracted, whilst in others it is speedily accomplished. Some fruits retain their seed long after they are ripe, as though it were necessary they should be thoroughly dried. Some capsular fruits project their seeds to a distance, by the elastic force with which their valves suddenly burst when thoroughly ripe. The Balsams (*Impatiens*) are a familiar instance of this, in which the effect is accelerated or suddenly stimulated by the slightest contact of the finger. The genus *Oxalis* has the seeds covered with an elastic arillus, which suddenly bursts after the capsules have opened, and turning the inside outwards projects the seed to a considerable distance.

(279.) *Peculiar Modes of Dissemination.* — The ordinary effect produced by moisture upon the valves of a seed-vessel is to keep them closed; but there are some remarkable exceptions to this law. In the *Onagrariæ,* which grow naturally in moist places, the valves open in moist weather, and the seeds are then scattered. There is a small annual cruciferous plant, called the Rose of Jericho (*Anastatica hierochuntina*), which grows in the driest deserts. When the seeds are ripe the plant withers and the branches coil together, so that the whole mass forms a sort of ball. As the root is very small and unbranched, it is easily torn up by the force of the wind, and the plant is then blown along the surface of the soil until it happens to arrive at some pool of water, when the branches imbibe moisture and unrol: the pericarps also burst and the seeds are disseminated in a spot where they are able to germinate.

(280.) *Hypocarpogean Fruits.* — There are some plants which possess the singular property of ripening their seed under the ground. In some of these the blossoms expand in the air, and then the pericarp is drawn down or forced underground by the incurvation of the pedicle, as in the *Antirrhinum Cymbalaria*, *Cyclamen*, &c. The *Trifolium subterraneum*, a small species of clover not uncommon in the sandy districts of England, has its flowers arranged four or five in a head: the end of the pedicel emits some succulent spinous processes, which soon harden, and the whole is gradually thrust under the surface of the soil, where the seeds ripen and germinate.

Some plants possess two distinct modes of flowering, the one aërial and the other subterranean; and these either perfect the fruit on both stems, as in the *Vicia amphicarpos;* or else that which is produced on the underground stems alone arrives at perfection, as in the *Arachis hypogœa*, or ground-nut.

(281.) *Preservation of Seeds.* — Notwithstanding the ample provision which is made for securing a superabundant crop of seeds, infinitely beyond the number of individuals destined to spring up from their dissemination, there is another circumstance to be noticed in their history, which most materially diminishes the chance of any species being extirpated. This is the property which seeds possess of resisting decomposition, and of retaining their vitality whenever they are placed under circumstances favourable to their preservation. Seeds are capable of being longer preserved in proportion as they have been more thoroughly matured; and hence it is advisable to allow them to remain for a certain time in the pericarp after they have been gathered, in order that they may more completely elaborate the provision there prepared for their use. When thoroughly mature many seeds may be preserved for a very great length of time, provided they are not exposed to the influences of those causes which determine their germination, viz:— a certain elevation of temperature, the presence of oxygen, and the influence of water. There are some however which very soon lose the faculty of germinating after they are ripe, though they may be preserved in a state fit for food for a long time. The seeds of coffee, for instance, will not germinate unless they are sown within the space of a few weeks after they have become ripe.

The fact that seeds retain their vitality for very many years is well authenticated. De Candolle tells us that a bag of seeds of the sensitive-plant gathered about sixty years ago, has regularly supplied the Paris garden with fresh plants every year since then. Young plants have been raised from seeds of a French-bean which were taken from the herbarium of Tournefort, where they must have lain for more than a century. These examples are remarkable exceptions to the more general rule, that seeds cannot be artificially preserved in a *living* state for many years together. It is certain that most of those found in ancient tombs, and in the catacombs of Egypt, have entirely lost their vitality; and although recent accounts have been published to the contrary, the fact does not seem to have been thoroughly established, and may possibly have been founded on some mistake, or perhaps imposition practised upon the credulity of the traveller by the cunning of the natives. M. Rifaud, a recent and laborious investigator of the antiquities and natural history of Egypt, brought to Europe a large collection of various seeds, bulbs, and other parts of plants, which he had found in the catacombs, and all of these were deprived of any vegetating power. Many of them have preserved to a great extent the appearance of freshness.

Some spikes of maize, obtained from the tombs of an ancient and extinct race in South America, still retain their original colours, the pericarps being either red or yellow; the variety is also much smaller, and in other respects different from those at present in cultivation. But although it is *generally* impossible to secure the vitality of seeds by artificial means for such very lengthened periods, it should seem that naturally and under peculiar circumstances, they can retain the power of germinating for many ages. It is very common, upon turning up the soil from great depths, or on breaking up a tract of ground which has lain uncultivated within the records of history, to find a crop of plants spring up from the newly-exposed surface, whose seeds must have lain dormant for centuries. In the fens of Cambridgeshire, after the surface has been drained and the soil ploughed, large crops of our mustards (*Sinapis arvensis* and *alba*) invariably spring up. Ray mentions the appearance of *Sisymbrium Irio* upon the walls of the houses immediately after the great fire of London, though the plant was not before known to exist in the neighbourhood. We must be cautious in not confounding such facts as we have here referred to, with the delusive effects sometimes produced upon soil which has been brought up from a great depth, and taken from strata which have never been disturbed before. The seeds of plants which spring up in such soils have been accidentally conveyed to them by the wind. We may also account for some cases where plants have appeared spontaneously on soils obtained from undisturbed strata at great depths, by supposing the seed to have been carried there by the percolation of water.

(282.) *Artificial Preservation of Seed.* — It is a vulgar notion that some seeds, as those of the melon and cucumber, improve by being kept for a few years; and that the plants raised from them will produce more fruit and fewer leaves than they would have done had they been sown immediately; but this opinion appears to be without sufficient foundation. In an economical point of view, the preservation of fruits and seeds in a state fitted for food is a subject of considerable importance; and various plans have been proposed which might combine both cheapness and the means of protecting them from the attacks of vermin, with security against decomposition. Some wheat preserved at Zurich for a space of 250 years was found to make excellent bread. One of the simplest and at the same time most efficacious modes of preserving corn, is to inclose it in wooden casks well pitched, and secured against the influences of the weather. When fleshy fruits are thoroughly ripe they become rotten, by the oxygen uniting with their carbon and forming carbonic acid. This effect may be prevented, and the fruit preserved for a considerable length of time in vessels hermetically sealed, and from which the air, or at least all the oxygen, has been previously expelled.

FIFTH PERIOD OF REPRODUCTION.

(283.) *Germination.* — When the maturation of the seed is complete, all further development of the embryo ceases, and it then enters into a state of torpidity; and thus it continues until it meets with that peculiar combination of circumstances upon which the last process of the general function of reproduction depends. After the dispersion of the seed has been secured, we might properly consider the function of reproduction to be terminated; but as the young plant is still dependent upon the nutriment previously provided for it, and has not yet acquired the power of preparing its own nutriment, we may perhaps be permitted to include the process of "germination," of which we are about to speak, among the details of the reproductive function. Germination commences with the revival of the embryo from its state of torpidity, and is considered to have terminated when the whole of the nutriment previously prepared has been absorbed, and the young plant is able to derive its nourishment in the usual way. This period bears some analogy to that of suckling in the Mammalia, or still more strikingly to that of incubation in birds.

(284.) *Stimulants to Germination.* — There are three requisites to germination, either of which being wanting the process will not take place. These are moisture, oxygen, and a certain elevation of temperature. When the conditions requisite for the germination of a seed are satisfied, it imbibes moisture through its integuments, the embryo swells, and the radicle is protruded and tends downwards. The plumule or terminal bud then expands and rises upwards; the albumen, either free or contained in the cotyledons, is soon exhausted; the young plant takes firm hold on the ground and commences its independent existence.

Although the period which elapses between the time when seeds are sown and when they first begin to germinate is very different even in the same species, according to the external conditions under which they are placed, yet if different seeds are subjected to precisely the same influences, we find a still more remarkable difference between the periods which elapse before they severally germinate. The following list exhibits the result of some experiments made at the Geneva garden, on seeds similarly watered and exposed to a common temperature of 9.5° R. It was ascertained that about half the species of the following families germinated after the lapse of the number of days here mentioned, viz:—

Days.	
9.	Amaranthaceæ.
10.	Cruciferæ.
11.	Cariophyllaceæ, Malvaceæ.
12.	Compositæ, Convolvulaceæ.
13.	Polygoneæ.
14.	Leguminosæ, Valerianeæ.
15.	Gramineæ, Labiatæ, Solaneæ.
20.	Ranunculaceæ.
22.	Onagrariæ.
23.	Umbelliferæ.

(285.) *Action of Moisture.*—It has been found that the quantity of water absorbed by seeds varies in proportion to their bulk, and that all seeds absorb very nearly a weight of water equal to their own. If a coloured liquid be used, it will be found to traverse the substance of the seed cover (*spermoderm*) until it collects in the cellular tissue near the extremity of the radicle. From this spot it is imbibed by the radicle, and penetrates into the cotyledons of dicotyledonous plants, along the minute and ramifying veins which traverse them. The chief use of the imbibed water appears to be, to dissolve whatever materials have been prepared in the seed for the nourishment of the embryo, and to convey them into its substance. Where the cotyledons are leaflike and not fleshy, they contain very little nutriment; and if there is no free albumen, the cotyledons themselves are furnished with stomata, immediately expand, and begin to elaborate nutriment by decomposing carbonic acid. When the alburnum is free and surrounds the cotyledons, it must in some way be absorbed by their surface, though it is difficult to explain how. The process bears a striking analogy to the suckling of the young in animals. Seeds will not germinate in boiled or distilled water, from which the oxygen has been expelled; and if they are placed in an atmosphere of hydrogen, azote, carbonic acid, or any gas which contains no portion of oxygen, they are equally incapable of germinating. They succeed best in a mixture of one part oxygen with three of azote, and this is not very far removed from the proportion in which these gases are united in the atmosphere. Where the oxygen is in larger quantity it over-stimulates the seed.

(286.) *Action of Oxygen.*—One use of oxygen in germination is to unite with the superfluous carbon which has been prepared during the process of maturation for the better preservation of the seed: thus it appears that the first step in the new process is to undo the last by which the maturation was completed. Consequently it is found that if the nearly ripe seed be sown immediately it is gathered, it will vegetate more speedily than when it has remained in the pericarp until the complete elaboration of the juices has taken place. This fact seems to account for the very rapid manner in which corn vegetates in moist and warm weather, after it has been cut and whilst still in the sheaf, or even before it is reaped.

(287.) *Action of Heat.*—The degree of heat requisite to produce germination is different for seeds of different species; but, within certain limits, an increased temperature acts as a stimulus upon all of them, the larger and drier seeds requiring a longer time for the effect to be produced.

(288.) *Action of Light.*—The action of light, though not fatal is decidedly noxious to the germination of seeds; and the cause why it is so is obvious. Seeds require to be freed from their superfluous carbon, by this combining with oxygen; but light is the chief stimulus which operates in the decomposition of carbonic acid, and in the fixation of carbon in the green parts.

(289.) *Action of the Soil.*—After germination is complete, most plants grow in some soil adapted to their nature, which serves them as a support, and more especially regulates the right proportion of moisture requisite for their roots.

(290.) *Vitality of the Embryo.*—Every part of the perfected embryo appears to be equally endowed with life; for if any portion be cut off, the remainder continues to germinate for a time, and will often reproduce the organ which has been detached. Thus the radicle may be repeatedly cut away whilst it is developing, and the plumule will nevertheless elongate; or the plumule may be cut away and the radicle will develop. There is of course a limit to these mutilations, beyond which the young plant cannot be made to grow; but whilst it is still germinating, the vital force cannot be said to reside in any one part of the individual rather than in another.

(291.) *Connection between Buds and Embryos.*—We

have already given several instances of the close affinity which subsists between the various foliaceous appendages on the stem (art. 85.), and have further mentioned the community of origin in the leaf-bud and flower-bud. There also exists an evident and striking affinity between the leaf-bud and the embryo, inasmuch as each of them when detached from the plant on which they were formed, is capable of becoming a perfect individual. The chief distinction between them consists in the former first developing its ascending organs and then its descending organs, whilst the embryo first emits the root and then develops the plumule.

(292.) *Proliferous Flowers.* — In "proliferous" flowers especially, the identity of their origin is strikingly exhibited. In these instances, every bud which in ordinary circumstances would have been developed as a flower, assumes the characters of a young plant. In the onion tribe this description of monstrosity is very common, and the little flowers which are aggregated into heads become small bulbs, and germinate as young plants even whilst they are still attached to the summit of the stem. The same fact very often takes place in certain grasses, and especially in some of those which affect a mountainous situation. This appears to be a provision of nature, to furnish an additional security against the chance of failure in the seed, at an elevation where the cold might offer a serious obstacle to its being perfected.

(293.) *Buds on Leaves.* — The *Bryum calycinum* furnishes one of the most satisfactory examples of the connection which exists between the bud and the embryo. Its leaves are very fleshy, and when they are placed in a moist situation, and even whilst they are still attached to the stem, little buds are formed at the bottom of the crenations on their margins (*fig.* 168.), and these buds soon develop into perfect plants. Now if we only suppose a leaf of this plant to be longitudinally folded inwards, and that its margins become grafted together, the buds will then correspond to the ovules arranged on the placenta of a carpel — an organ which we have considered to be formed on this principle (art. 100.).

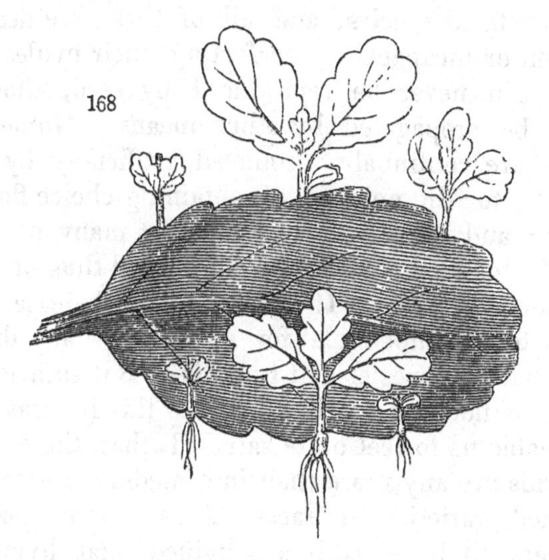

(294.) *Proportion between Seeds and Buds.* — An argument in favour of the common origin of the embryo and bud is deduced from the observed fact, that many plants which produce the one in abundance are proportionally defective in the other kind. But this after all may depend upon the plant not being able to provide a sufficiency of nutriment for both.

(295.) *Hybrids.* — If the pollen of one species is employed to fertilise the ovules of another, the seeds will often produce plants which are strictly intermediate in all respects between the two parents. Such productions are termed hybrids, and are manifestly analogous to mules among animals. The conditions necessary for the production of a hybrid are not ascertained, beyond the fact that those species only are capable of forming them which are nearly allied to each other, and are either of the same genus, or of genera which scarcely differ. It has been suggested that the possibility of producing hybrids was limited to species whose pollen, or rather whose pollen granules, were nearly of the same form and dimensions; but this is at present mere conjecture. Not more than forty kinds of hybrids have been found *naturally* produced in a wild state between well-defined species, and all of these are described as barren or incapable of perfecting their ovules; so that they can never be reproduced by seed, though they may be propagated by other means. Numerous hybrids are continually produced artificially by horticulturists, for the purpose of obtaining choice flowers and fruit; and it has been asserted that many of these are capable of fertilising their ovules, and thus of being reproduced by seed. If this be really the case, it would seem to be impossible for us to draw any distinction between true species and hybrids. But sufficient attention has not hitherto been paid to this intricate subject, to enable us to feel quite satisfied that these supposed hybrids are any more than intermediate forms between marked varieties or races of the same species. It appears to have been ascertained that hybrids may be fertilised by the pollen taken from one or other of the parent species, and that the seed thus obtained will produce plants intermediate between that species and the hybrid, and thus a return may gradually be made to one of the original types. It has been equally asserted of animals, that although mules never produce young between themselves, yet a female mule may become productive by a male of one or other of the parent species.

The rarity of wild hybrids is easily accounted for by the fact, that so soon as the stigma has been affected by the contact of the pollen, it becomes incapable of transmitting an additional influence from any fresh grains that may afterwards be applied to it; and consequently the chances of every stigma being first affected by the pollen of its own stamens (if we except diœcious species), is infinitely greater than its receiving any influence from others.

(296.) *Permanence of Species.*—Every thing that has hitherto been written on the origin and limitation of species, may be fairly stated as purely hypothetical. Linnæus supposed that only a few species or distinct typical forms were originally created, and that a multitude of others had since been derived from them by repeated intermixture and crossings. He supposed the species of very different genera might be capable of intermixing and producing new species, and even new genera. These speculations are wholly unsupported by facts or experiments. De Candolle also supposes a definite number of species or typical forms to have been originally created, but he does not imagine any decidedly new form or type to have ever originated from them. He considers that certain hybrids can reproduce their kind, but that in such cases there exists a constant tendency in the offspring to return again into one or other of the original types from which they sprang. Thus we should never have any strictly new type introduced, or any form which differed very materially from what was already in existence, but only a multitude of minute shades of difference, in varieties which were all intermediate between the original species. In this way he proposes to account for the endless varieties of some of our long cultivated fruits, as apples, pears, &c. The subject is one of great difficulty, and it will require many accurate and careful experiments to be made, before we can expect to ascertain the laws by which the limitation of species and the production of hybrids are regulated. We are quite certain that many forms, considered characteristic of particular species, have continued unaltered in their minutest particulars for the last 3000 years at least. This is proved by a careful examination of the fragments of numerous plants found in the catacombs of Egypt. An analogous fact is still more strikingly established in the animal kingdom, and for a much longer period; since the forms of certain existing species of shells have been found in those tertiary deposits of which the geologist can say no more than that they are comparatively recent in the history of our globe, though incalculably earlier than any date to which we can refer by authentic records.

(297.) *Origin of Varieties.*—The origin of varieties is a phenomenon in some respects analogous to the creation of hybrids; and it has been even supposed that all races, or such varieties as are capable of maintaining their peculiarities by seed, must have originated in hybridity between two species. If such hybrids have been fertilised by the parent species, and new hybrids of the second and third degree been produced, these will so closely resemble the parent plants that they will appear to be mere varieties of it.

CHAP. VII.

EPIRRHEOLOGY, BOTANICAL GEOGRAPHY, FOSSIL BOTANY.

EPIRRHEOLOGY (298.). — DIRECTION OF ROOTS AND STEMS (299.). — BOTANICAL GEOGRAPHY (302.). — FOSSIL BOTANY. (315.).

(298.) *Epirrheology.* — This term has recently been proposed, to express that branch of our science which treats of the effects produced by external agents upon the living plant. It can only be considered as a subordinate department of vegetable physiology, and one indeed whose limits are not very strictly defined. For we have seen that life itself requires the stimulus of external agency, in order that its powers may be elicited, and produce the various phenomena of vegetation included under one or other of the two functions of nutrition and reproduction. But then these functions become variously modified, according as the external stimuli by which they are called into action are permitted to operate with greater or less intensity. In all cases, there is that happy mean which can so regulate the vital force as to produce a healthy and vigorous condition of existence; whilst every increase or diminution in the stimulus applied, only tends to injure or greatly to modify the individual subjected to its long-continued influence. Physiology might be considered as embracing the investigation only of such phenomena as resulted from the healthy condition of the vital functions; whilst epirrheology would take further cognisance of such as resulted from an unhealthy condition of vegetation. Hence this department would lay the foundations of another branch, termed the "nosology" of plants, or that science which treats of their diseases; and also of the extensive subject of "Botanical Geography," which makes inquiry into those causes which limit the distribution of various species to certain spots upon the earth's surface. But in a treatise like the present we have not thought it necessary to make any distinction between physiology and epirrheology, nor are we prepared to allow that such distinction is a very judicious one. In order to understand the effects produced by the vital force, it is necessary to trace its operations under various modifications of the external stimuli by which it is controlled, and even rendered capable of acting at all. These inquiries relate to the results of an action and reaction between opposing forces, questions which cannot well be separated without greater refinement than the subject seems to require. There are, however, certain phenomena, the discussion of which could not be conveniently introduced under either of the two functions into which the vital properties were arranged. Of these we may select as an example the effects produced by the action of gravity upon growing plants.

(299.) *Direction of Roots and Stems*.—That the roots and stems of most plants constantly develop in opposite directions, is a fact too notorious to need a comment; and any deviation from this general law is considered an anomalous circumstance. It is not strictly true to say that the tendency of all stems is upwards, though it is more nearly true that all roots take a direction downwards. The branches of the weeping birch, weeping willow, and some others of this character incline downwards, merely by the effect of gravity, acting upon the long slender rods of which they are formed. But there are some trees, as the weeping ash, and weeping horse-chestnut, whose branches take a decidedly downward tendency from their very origin. Many plants also have underground stems (*rhizomata*), besides those which they develop above ground. But, neglecting these anomalies, it is generally true that the stem has a tendency to develop upwards, and the root downwards. There are two causes to which we may ascribe these modifications in the directions of the stems and roots. One is "gravity," and the other "light."

(300.) *Effects of Gravity on Vegetation.*—That gravity is an important agent in determining the difference between the directions taken by the root and stem, is shown by an ingenious experiment of Mr. Knight. He placed some French-beans on the circumference of two wheels, and so secured them that they could not be thrown off when a rapid rotatory motion was given to the wheels. One wheel was disposed horizontally, and the other vertically, and both were kept in constant motion whilst the beans were germinating. The radicles of those beans which germinated on the vertical wheel extended themselves outwards or from the centre, and the plumules inwards or towards it. Those which were placed on the horizontal wheel pushed their radicles downwards and their plumules upwards; but the former were also inclined from, and the latter towards the axis of the wheel. This inclination was found to be greater in proportion as the velocity of the wheel was increased. Now in the vertical wheel the effects of gravity were nullified, since the beans were constantly changing their position with respect to those parts which were alternately uppermost and lowermost in each revolution. The only cause which could have produced the effects described must be the centrifugal force, which has here replaced the force of gravity, compelling the root to grow outwards and the stem inwards, instead of downwards and upwards. The effect produced upon the horizontal wheel is evidently the result of the combined action of the two forces — gravity inclining the root downwards, and the centrifugal force propelling it outwards; and the reverse with regard to the stem. Although it is plain that gravity is the efficient cause in establishing the directions of the stems and roots of plants, it is not so easy to understand the manner in which it produces opposite effects on these two organs. Various theories have been formed to account for this, and the most plausible is that which ascribes it to the different manners in which the newly developed tissues are added to the root and stem. In the root the addition is almost entirely confined to the very extremity, whilst the stem continues to increase for some time throughout its whole length. Hence it is supposed that the soft materials continually deposited at the extremity of the root must ever be tending downwards from the mere effect of gravity alone. In the stem, gravity would cause a subsidence of the denser and more nutritious materials to the lower side, and this side would consequently be more nourished than the upper, supposing the stem to be somewhat inclined from the perpendicular. The consequence of one side being better nourished than the other, whilst the whole was in a growing state, would be a greater extension of that side; and thus a slight curvature upwards would be given to the stem, which, being continually repeated as it develops, would always tend to keep it more or less in a vertical position. Perhaps we want sufficient data to allow us to lay any great stress upon this explanation.

(301.) *Effect of Light on Vegetation.* — Light is another cause which produces a great effect in modifying the directions of the stems of plants. When grown in a chamber which admits the light on one side only, they constantly incline towards it. This has been supposed to be owing to a greater decomposition of carbonic acid on the side which is towards the light, and a necessarily greater deposition of carbon on that side than on the other. This produces a greater rigidity in those parts, and consequently a curvature on the side which is towards the light. This effect is produced only on the young green parts of plants, and does not take place in the old woody portions; nor is it observed in parasitic species, which are without the means of decomposing carbonic acid. The missletoe forms a most remarkable exception to the usual laws which regulate the direction of the root and stem. If a seed of this plant be attached to a piece of glass placed over a dark surface, the radicle invariably extends itself in a direction opposite to the side in which the light shines, from whatever quarter it may come. The branches of this plant are also developed indifferently in all directions, without any obvious tendency either upwards or towards the side from whence the greatest illumination may proceed.

(302.) *Botanical Geography.* — We cannot dismiss the physiological department of our subject, without referring to that branch of it which treats of the natural distribution of plants on the earth's surface — in other words, to "Botanical Geography." It is a fact sufficiently familiar to every one, that different species of plants affect peculiar situations; some love an exposed aspect, others prefer shady places; some are found in mountainous districts, others in plains, in marshes, and even wholly submerged in lakes, or in the sea. The various physical circumstances attending different spots in the same range of country determine the "stations" in which the different species of plants can grow. We know that different plants require different degrees of temperature; some are calculated to live in cold or temperate climates, whilst there are others which belong to the torrid zone;

and these last we are obliged in our latitudes to preserve in the stove or conservatory. The term "habitation" has been given to any tract of country throughout which each particular species is found naturally distributed in stations adapted to its growth. The determination of these stations and habitations of plants leads to an inquiry into the laws and circumstances which regulate the distribution of species. We must suppose that there exists a mutual relation between the external conditions under which each species is naturally disposed, and its own peculiar organization; and this relation must be sought for by a patient comparison of the various species, genera, and families peculiar to different regions, with the precise conditions under which they there exist. The problem is one of a most complicated description, and it cannot be said that any very decided progress has hitherto been made towards its solution. We shall mention some of the more obvious conditions under which all inquiries of this description must be regulated, and present the reader with some of the conclusions at which botanists have already arrived.

Influence of external Circumstances on the Geographic Distribution of Plants.

(303.) *Temperature.* — The influence of temperature is the most decided of all the circumstances which regulate the distribution of plants on the surface of the earth. It seems evident, that each species is *constitutionally* adapted to thrive best between certain limits of temperature, and that every excess of heat or cold (beyond these) is alike injurious to it. Hence, every species must naturally be restricted within those geographical boundaries beyond which the temperature either exceeds or falls short of these limits. These boundaries will not necessarily coincide with any definite parallels of latitude; for it is well known that the climate of different places having the same latitude is very different. By drawing lines through those places where the mean annual temperature is found to be the same, Humboldt established a series of "Isothermal" lines intersecting the parallels of latitude. But these lines by no means show us what might be the probable range of particular species. For an isothermal line may intersect a range of country where the extremes of heat and cold are very different; and the constitution of different species, which may be equally adapted to a given mean temperature, may not be equally suited to these differences in the extremes. Thus many plants which will live in the open air at Edinburgh, would perish during the severer winters of more southerly regions, whilst many that can stand greater cold than that to which they would be exposed at Edinburgh, require also greater heat in the summer than they would find there, in order to bring their fruit to perfection, or even to ripen their wood sufficiently to maintain them in a healthy condition. In fact, the *mean* distribution of temperature throughout the year, is a consideration of much less importance than the distribution per month, which perhaps most effectually regulates the range of species. As annuals cannot maintain their footing in any climate without yearly perfecting their seeds, they are necessarily limited to more temperate habitations than certain perennials; it is sufficient for the latter, if they occasionally meet with a season in which they may be able to do so. It has been remarked that the western parts of continents are more nearly equable in their temperature throughout the year than the eastern, and the southern hemisphere than the northern; and that evergreens affect the former, and deciduous trees the latter description of climate. Maritime districts have always a more nearly equable temperature than such as are inland.

Besides the physiological relations which plants possess with regard to temperature, there are others of a physical character by which their distribution is considerably affected. Where the temperature is so low that water exists only in the form of ice, it cannot be imbibed by the roots, and no plants can live there.

When the sap is frozen, the cells and vessels in which it is contained are ruptured, and the parts subjected to such an accident die. But trees possess a resource against the effects of great cold, in their roots penetrating to a depth beyond that which the frost has reached. Hence they obtain a supply of caloric, which is not readily carried off, because their woody layers and bark are bad conductors of heat. It has been observed that the internal parts of large trees retain a temperature which is about equal to that of the subsoil at one half the depth of their roots.

The temperature of a tree, being always influenced by that of the subsoil, will be greater than the surrounding atmosphere during winter in high latitudes, and less during summer in low latitudes. This is even more remarkably the case than would at first be imagined, if we were to refer the cooling and heating of the earth to the effect of radiation alone. But it has been observed by Von Buch, that the temperature of the subsoil is mainly regulated by that of the surface waters, which by infiltrating into the earth produce an effect far greater than any which may be ascribed to the mere conducting power of rocks and soils. Now, in the frigid zone, no infiltration takes place during the winter, when every drop of water is converted into ice or snow; and consequently the mean temperature of the subsoil in very high latitudes, will be somewhat higher than the mean temperature of the atmosphere; but this is not so in lower latitudes, where the infiltration continues during a great portion of the winter. On the other hand, as we approach the torrid zone, where rain falls only during the coolest season of the year, the mean temperature of the subsoil will be more cooled in proportion than in those places where it also falls during the hottest weather. Hence it happens that the mean temperature of springs throughout the central and northern parts of Europe, as far as Edinburgh, are much the same as the mean temperature of the air; whilst from the south of Europe to the tropic of Cancer, the difference is gradually increasing in favour of the atm os

But from the latitude of Edinburgh northwards, the difference increases in favour of the subsoil. The consequence is that certain plants which naturally belong to the more temperate parts of our zone, are enabled to extend themselves further north and south than they could do if the mean temperature of the soil and air were every where the same.

(304.) *Influence of Light.* — The influence of light is less essential than that of temperature in fixing the geographical limits of different species, though it is certainly of great importance in many cases. Light is, as we have seen, the chief agent in stimulating the vital properties, and its effects are apparent in a great number of vital phenomena, such as the absorption of the sap, the exhalation of moisture, and the decomposition of carbonic acid. It is probable that each species requires a peculiar stimulus from different degrees of light as well as of heat, and we find that such as are succulent, resinous, or oily, generally prefer situations where they can obtain most light; whilst many evergreens and others grow best where they are somewhat shaded. In these respects alpine plants may be contrasted with maritime species. the former receiving the greatest and the latter the least light, under the same degree of latitude. Whilst the mean distribution of light is more nearly equable for all latitudes than the mean temperature, the variations in the mode of its distribution are much greater. Contrast, for instance, the alternate long continuance of light and darkness at the poles with their nearly equable daily distribution at the equator.

(305.) *Influence of Moisture.* — The proportion in which water is supplied, constitutes one of the chief peculiarities of every "station;" and plants are very differently constituted with respect to the precise supply which they require to preserve them in a healthy condition. Those which require most, have a loose and spongy texture, with large and soft leaves, and little or no pubescence, but many stomata; whilst such as grow in arid districts are frequently firm and succulent, often provided with long pubescence, but have few stomata. An excess of water is apt to corrupt and dissolve the outer texture, and hence we find many aquatics, as the pondweeds (*Potamogeton*), protected by a superficial varnish. Many Monocotyledons are coated with a siliceous pellicle, and afford useful materials for thatching, as the common reed.

(306.) *Influence of Soils.* — Most soils are a very heterogeneous mixture of different earths and other matters; and hence it is not likely that any very decided feature will be often impressed upon the flora of a given district, by any peculiarity in the purely chemical qualities which soils possess. That some chemical action takes place in certain soils cannot be positively denied, but has probably been greatly exaggerated. For though certain plants seem to prefer particular geological districts marked by the prevalence of peculiar rocks, some especially abounding on limestone and chalk, others on slate-rock; yet it not unfrequently happens that many of these plants also occur in equal abundance in some other localities where the rocks possess a totally different mineralogical character. It seems, therefore, more likely that such effects may be attributed to mechanical rather than to chemical causes; especially to the mode in which different rocks disintegrate, and are rendered capable of retaining a greater or less abundant supply of moisture. It may indeed be said, that these mechanical properties are generally the direct result of the peculiar chemical qualities which the rocks possess, though in some cases rocks of very different mineralogical character certainly disintegrate in much the same manner. Hence we find the same lichens and some other plants growing on schistose rocks, whether they happen to be argillaceous or cretaceous in their composition. Various soils may be stated as generally retaining moisture in proportion to the quantity of alumina which they contain, and parting with it more readily in proportion as they abound in silica. Siliceous tracts require most rain, and clay soils least, to become proportionably fertile. Sandy districts support only such low or trailing plants as the wind cannot readily root out, or those which have very deep and branching roots; whilst very tenacious clays are adapted only to such species as have small roots, and which do not require any great depth of earth.

(307.) *Influence of the Atmosphere.* — Although the atmosphere is every where of the same chemical composition, its effects may vary in proportion to the density which it possesses at different elevations, or according to the materials (as moisture, gases, &c.) which may be suspended in it; or lastly according to its mechanical action, in the greater or less degree of violence with which it is moved in different regions. It is probable that the difference in density which the atmosphere possesses at different elevations above the surface of the earth, produces little or no effect in comparison with those which result from the modifications which the temperature, light, humidity of the air, &c. undergo. Since the mean temperature diminishes in receding from the equator much in the same proportion as in ascending a mountain, many plants peculiar to the plains of higher latitudes are found on the tops of mountains in warmer climates. Hence a very extensive range may be given artificially to some plants, by cultivating them at different altitudes in different latitudes. Humboldt has likened the earth to two great mountains whose bases meet at the equator, and whose summits are the poles; and, *ceteris paribus,* we may say that the latitude at which a plant thrives best will vary as the altitude above the sea at which it also flourishes under the tropics. The potato offers an interesting illustration of this fact — growing in Chili, at an altitude of eleven or twelve thousand feet above the level of the sea, and being well adapted to summer culture in the plains of the temperate zone as far north as Scotland. The olive has a much less extended range, and can only be cultivated as far north as 24°, and at an altitude of twelve hundred feet in tropical climates.

(308.) *Botanical Stations.*—The various peculiarities which characterize different "stations," can scarcely be appreciated. Those which possess a very general resemblance, may still differ in some important circumstance by which the existence, or at least the prevalence of some peculiar species may be determined. Thus a marsh may be formed by salt and fresh water mixed in different proportions; two tracts in other respects alike, may be very differently exposed to the prevalence of winds, or the influence of sea breezes, &c. Independently of these modifying circumstances, we may enumerate about sixteen tolerably well defined stations, to one or other of which the different plants of every flora will be found more particularly attached.

1. *Maritime.*—Districts bordering on the sea and influenced by the spray and sea breezes.

2. *Marine.*—Where plants are growing beneath or on the surface of the sea itself.

3. *Aquatic.* — Freshwater rivers and lakes, where the plants are wholly immersed or floating on the surface.

4. *Marsh.*—Bogs and fens.

5. *Meadows and Pastures.*

6. *Cultivated Lands.*—These districts abound in plants which have been introduced by the agency of man, and have become completely or partially naturalized.

7. *Rocks.* — Lichens, mosses, and other cryptogamic tribes abound in rocky situations, but more especially in the vicinity of springs and cascades. A few phanerogamic plants also affect such situations, even where there is little or no soil to support them.

8. *Sands.*

9. *Barren Tracts,* by road sides, &c.

10. *Rubbish.*—There are many species which seem to follow the footsteps of man, and spring up wherever he scatters the rubbish and rejectamenta of his dwellings.

11. *Forests.*—These districts may be considered with respect to the trees which compose the forests, and also with reference to the humbler species which seek their shade.

12. *Copses and Hedges.*

13. *Subterranean Caves.*

14. *Alpine.*

15. PARASITIC. (See art. 234.)

16. PSEUDO-PARASITIC. (See art. 234.)

(309.) *Botanical Habitations.*—Greater uncertainty prevails respecting the different habitations of plants than their stations. If indeed the extent of their habitations were entirely dependent upon their range in latitude, the difficulty of determining them would not be so great; but it is a remarkable circumstance, that the vast majority of species grow naturally within certain limits restricted in longitude as well as in latitude; that is to say, the limits within which they naturally occur, are much more restricted than the regions throughout which they might readily grow, so far as climate is concerned in this question. There are indeed some species which have a very extensive range in longitude as well as in latitude, and are even found in both hemispheres, but several of these have undoubtedly become thus generally dispersed by the agency of man. Others we may equally conclude to have been transported by natural causes, from the habitations to which they were first restricted. But when we have made all such allowances, we find the great majority of species so far restricted in their range, as to lead us to the very probable supposition that each was originally assigned by the Creator to some definite spot upon the surface of the earth, from whence it has wandered to a greater or less extent in all directions, until it happened to meet with such obstacles as were sufficient to check its further progress. It may be worth while to consider the nature of those obstacles which afford the most effectual barrier to the migration of species from one part of the earth's surface to another; and also the means by which their migration is most effectually provided for.

(310.) *Obstacles to Migration.*—

1. *Seas.*—The salt of sea-water produces an injurious effect upon the seeds of plants, and completely destroys the vitality of those which are long subjected to its influence. In proportion therefore to the extent of sea which surrounds a tract of land, the chances are diminished by which the seeds of plants may be wafted to or from it in a state fitted for germination. This is remarkably exemplified in the flora of St. Helena, which is so peculiar, that not more than two or three of its indigenous species have been found on the continent of America, and not one of them on the continent of Africa. Generally speaking, the floras of all islands resemble those of the continents to which they are nearest, in proportion to their greater proximity to those continents. England does not possess fifty species which have not also been detected in France; and probably, the number peculiar to our flora is even still less than this. The floras of the opposite shores of the Mediterranean are very nearly the same.

2. *Deserts.*—These are a very effectual barrier to the migration of species; and hence there are scarcely any species described in the "Flora Atlantica" which are to be met with in Senegal; the great desert of Sahara completely intercepting the botanical intercommunication of the two districts.

3. *Mountain Chains.* — Where mountain chains possess lofty summits, the cold of those regions presents a barrier to the migration of plants across them. In general however they are not so effectual as seas and deserts, on account of their being intersected by transverse valleys.

4. *Partial Obstacles* are offered by extensive forests and marshes; for although there are numerous species which prefer such tracts as "stations," to which they are best adapted, there are others which cannot live

under the influence of the moisture and shade which prevail there.

(311.) *Means of Transport.*—

1. *Currents.*—Rivers and other currents of fresh water are among the most effectual means of dispersing the seeds of plants: even the sea may occasionally serve a like purpose where the seed is protected from its influence by some accidental circumstance.

2. *Atmosphere.*—Many seeds are provided with downy and winglike appendages, by which their dispersion is secured; but more especially the minute impalpable sporules of cryptogamic plants appear capable of being wafted to very considerable distances by this means. It has been supposed that two species of lichen found on the coasts of Bretagne, have been brought thither from Jamaica by the prevalence of the south-west winds.

3. *Animals.*—Seeds often become entangled in the hair and wool of many animals, and may thus be carried by them to considerable distances from the spot where they grew; but more especially such as are furnished with hard pericarps, or bony coverings to the kernel (as in stone-fruits) are capable of resisting the digestive powers of the stomach, and are thus conveyed by birds from one region to another in a state fitted for germination. But man is most instrumental in the dispersion of different kinds of plants. The seeds of some he has carried intentionally from one quarter of the globe to another; and others have been accidentally transported by him in a thousand ways, and follow his footsteps wherever he has penetrated.

(312.) *Botanical Regions.*—It seems to be a natural consequence of our considering the geographical distribution of every species to have taken place by its gradual dispersion from one definite spot on the earth's surface, that some would be found only in one district, and others in another, provided these were separated by some great physical feature, such as a chain of mountains or a wide sea; and that two such districts, though they might lie under the same parallel of latitude, would contain few species common to both. Such districts are termed "botanical regions." These are spaces enclosing particular species, distributed through them in the stations adapted to their growth; but so encompassed by physical obstructions, that the great majority of species found within their limits are not to be met with elsewhere. We do not as yet possess any very accurate information respecting the number and exact extent of the well-defined botanical regions into which the surface of the earth may be mapped out. There are about fifty whose floras have been partially examined, and of which the following list has been given:—

1. *Arctic.* — Includes the northern parts of Asia, Europe, and America. This region is not well defined towards the south; but may be considered as terminating in that direction between lat. 62° and 66°.

2. *European.*—Included within a line drawn from the north of Scotland, through St. Petersburg, the Ural Mountains, to the north of the coasts of the Mediterranean up to the Pyrenees.

3. *Mediterranean.* — Coasts all round the Mediterranean, with Italy, Dalmatia, Greece, Syria, and Spain.

4. *Red Sea.*—Includes Egypt, Abyssinia, and part of Arabia.

5. *Persian.* — Includes countries round the Persian Gulf.

6. *Caucasian.* — Caucasian chain and countries between the Euxine and Caspian.

7. *Tartarian.* — About Lake Aral.

8. *Siberian.* — Between the Northern Ocean and the Ural Mountains. Bounded towards the south by the Altaic Mountains.

9. *Nepaul.* — The chain of the Himalaya.

10. *Bengal.*—The plains through which the Ganges flows.

11. *Indian.* — The Peninsula and Ceylon.

12. *Birman empire.*

13. *Cochin-China.*

14. *Indian Archipelago.*

15. *New Holland,* with Van Diemen, New Zealand, New Caledonia, Norfolk Island.

16. *Friendly and Society Islands,* with those adjacent.

17. *Sandwich Islands.*

18. *Mulgrave, Carolinas, Marian,* &c.

19. *Philippine Islands.*

20. *China,* with Corea and Japan. Too little known to be subdivided.

21. *Aleutian Islands,* and the north-west of America.

22. *North-east of America.*— Canada and the United States.

23. *Mexico.* — From California and Texas to the Isthmus of Panama.

24. *Antilles.*

25. *Venezuela,* Carthagena, and the Oronoco.

26. *New Grenada* and Quito.—Includes every variety of climate, from the sea-shore to the summits of the highest Andes.

27. *Guyana.* — Cayenne, Surinam.

28. *Peru.*

29. *Bolivia.*

30. The *Basin* of the Amazon.

31. *North-east of Brazil.*

32. *South-east of Brazil.*

33. *West of Brazil.*

34. *Argentine Region.* — Between the Andes of Chili, Paraguay, Brazil, and Patagonia.

35. *Chili,* with the Island of Chiloe.

36. *Patagonia.*

37. *Ascension, and St. Helena.*

38. *Tristan d'Acunha, and Diego d'Alvares.*

39. *Prince Edward's, Marion, Kerguelen, and St. Paul.*

40. *Cape of Good Hope,* with all extra-tropical Southern Africa.

41. *Madagascar,* with the Mauritius and Isle of Bourbon.

42. *Congo.*

43. *Guinea.*

44. *Senegambia.*

45. *Canaries,* Madeira and Azores.

The centres of Africa, Asia, and other unexplored districts probably afford several more regions.

Twelve of the regions enumerated belong to the northern hemisphere, between the pole and tropic of Cancer; twenty-six are intra-tropical; and seven are extra-tropical, in the southern hemisphere. The first are the largest, and approach each other the nearest; the second are less extended, and more frequently separated by the ocean and deserts; the last are very unequal in extent, and above all more dispersed, many of them being small islands in the midst of an immense ocean.

(313.) *Relative Number of Individuals and Groups in each Region.* — In contrasting one botanical region with another, inquiry may be made as to the number of individuals which each may be supposed to contain, and also as to the number of species, genera, and families. The result of the first of these inquiries must depend upon the actual extent of country included in the region, and upon the character of its climate. The nature of the plants which grow in the region will also form an important element in this inquiry, since a space occupied by a single tree may contain many hundreds of smaller plants, and those regions in which large species prevail will not contain so many individuals as those which abound in small ones. The greater or less prevalence of particular species in a given region, may be observed by noting the number of places in which they occur; and then representing by ciphers the relative abundance in which they appear to exist in each spot, the sums of these ciphers will afford some approximation to the relative abundance of each species. Those regions which embrace a greater diversity of stations will, *ceteris paribus*, also contain a greater number of species. Those which are more strictly isolated from each other are not so likely to interchange their species; and hence it is observed, that a given space on a continent generally contains a far greater number of species than an equal space in an island. An elevation of temperature is favourable to the greater number of species, as we find by the fact that the number at different latitudes increases as we approach the equator. The genera and families also seem to obey a similar law; but we scarcely possess sufficient information to speak positively as to the proportion in which the relative rate of their increase takes place. It does not appear that the same proportion of genera to species is maintained in different latitudes: for instance, the species in Sweden are to those in France as one to three; whilst the genera are as one to two.

(314.) *Proportion of Species in each Class, in different Regions.*—If a botanist collect indiscriminately all the plants he meets with, in any region he may be examining, he will most probably be soon able to obtain a very close approximation to the relative proportion which the species of each of the three classes, and many of the orders bear to each other, long before he has obtained an accurate notion of the whole number of species which the region possesses. So far as calculations have hiterto been made, the following general laws appear to be correct; and it is not likely that they will be modified by any additional information which future researches may procure.

1. The proportion of cryptogamic to phanerogamic species increases as we recede from the equator.

2. The proportion of Dicotyledones to Monocotyledones increases as we approach the equator.

3. The absolute number of species, and also the proportion of woody species to the herbaceous, increases as we approach the equator.

4. The number of species either annual or biennial

(*monocarpeans*) is greatest in temperate regions, and diminishes both towards the equator and poles.

Many local circumstances produce remarkable modifications in the relative proportions between the species of different classes and orders, in regions under the same parallels of latitude. Thus for instance, *ceteris paribus*, the cryptogamic tribes flourish most in moist regions. The places best adapted to the growth of ferns are the islands in tropical climates, in some of which, as in St. Helena, one half the flora is composed of them. It is remarkable that in this respect, and as regards the existence of arborescent species in this order, the ancient flora of our coal-fields, appears to approximate very closely to that of islands situate in the midst of an extended ocean and in low latitudes. The same causes which appear favourable to the increase of cryptogamic species, seem also to produce a diminution in the proportions which dicotyledons bear to monocotyledons. Other relations of considerable interest have been pointed out between the species of different orders, occurring in different regions; but we cannot enter into the minutiæ of their details, our object being rather to present the reader with the principles on which such investigations depend, than to acquaint him with the partial results which have hitherto been deduced from them; several of which must doubtless be greatly modified hereafter, considering the little knowledge we at present possess of the floras of many parts of the world.

The following table exhibits a few of those results which appear to have been most satisfactorily established. It gives the relative proportion which ten well-defined orders, or families of plants, bear to the whole of the phanerogamic tribes in the torrid, temperate, and frigid zones respectively, and shows us in which they occur in the greatest *relative* abundance, decreasing as we recede from that zone towards the others.

Orders.	Equatorial. Lat. 0—10°	Temperate. 45°—52°	Frigid. 67°—72°	Maximum ratio in
Juncèæ - (*Rushes*)	$\frac{1}{400}$	$\frac{1}{90}$	$\frac{1}{25}$	Frigid.
Cyperaceæ (*Sedges*)	$\frac{1}{22}$ Old World $\frac{1}{50}$ New World	$\frac{1}{20}$	$\frac{1}{9}$	Frigid.
Gramineæ (*Grasses*)	$\frac{1}{14}$	$\frac{1}{12}$	$\frac{1}{10}$	Frigid.
Compositæ	$\frac{1}{18}$ Old World $\frac{1}{12}$ New World	$\frac{1}{8}$ $\frac{1}{6}$	$\frac{1}{13}$	Temperate.
Leguminosæ	$\frac{1}{10}$	$\frac{1}{18}$	$\frac{1}{35}$	Equatorial.
Rubiaceæ	$\frac{1}{14}$ Old World $\frac{1}{25}$ New World	$\frac{1}{60}$	$\frac{1}{80}$	Equatorial.
Euphorbiaceæ	$\frac{1}{32}$	$\frac{1}{80}$	$\frac{1}{500}$	Equatorial.
Malvaceæ -	$\frac{1}{35}$	$\frac{1}{200}$	—	Equatorial.
Umbelliferæ	$\frac{1}{500}$	$\frac{1}{40}$	$\frac{1}{60}$	Temperate.
Cruciferæ -	$\frac{1}{800}$	$\frac{1}{18}$ Europe $\frac{1}{60}$ Amer.	$\frac{1}{24}$	Temperate.

(315.) *Fossil Botany.* — The history of vegetation could not be completed without some inquiry respecting those plants which existed on the earth in its primæval state, during the extended geological epochs which elapsed before the establishment of the present order of things. Traces of this ancient vegetation are very abundant in certain strata, but more especially in the "coal-measures," the important mineral combustible obtained from them being nothing else than vegetable matter in an altered and fossilized state. In general, we do not find the remains of plants so perfectly preserved as the skeletons of vertebrate animals, or the testaceous coverings of mollusca. It is also rare to meet with those parts (the flower and seeds) upon which the distinction of species and their classification chiefly depend: but still the fragments which remain often possess very great beauty; and many specimens of wood are so exactly preserved, that their tissue may be distinguished under a microscope as completely as in recent species. As it is principally from these fragments of stems, and the impressions of leaves, that any comparison between the ancient and present flora of our planet must be instituted, it will be evident that such data must generally be far too imperfect to admit of any accurate determination of specific differences, though they may afford us sufficient materials for ascertaining several truths of high interest. The class, order, sometimes the precise genus, may be ascertained to which a fossil vegetable belongs, even though we posses only a small fragment of the plant. More frequently, these fossils bear an analogy to some recent genera, which they closely resemble, but to which they cannot be accurately referred. In such cases this resemblance is indicated by referring them provisionally to a genus whose name is a modification of the recent genus: thus "*Lycopodites*" is a genus of fossil plants allied to "*Lycopodium*," but too imperfectly known to have its characters fully pointed out.

(316.) *Botanical Epochs.* — It was soon remarked, when the study of fossil vegetables began to attract the attention of botanists, that those from the coal-measures were distinct from the plants now existing on the surface of the earth, and that they more nearly resembled the species of tropical climates than such as grew in the temperate zones. Subsequent researches have shown that the species embedded in different strata likewise differ from each other, and that on the whole there are about fourteen distinct gealogical formations in which traces of vegetables occur. According to Mons. Brongniart they first appear in the schists and limestones below the coal. These contain a few cryptogamic species (about thirteen), of which four are marine Algæ, and the rest ferns, or the allied orders. In the coal itself above 300 distinct species have been recognised, among which those of the higher tribes of cryptogamic plants are the most abundant, amounting to about two thirds of the whole. Many of them are arborescent, and parts of their trunks are found standing vertically in the spots where they grew. There are no marine plants in the formation. A few palms and

Gramineæ are the chief Monocotyledones; and there are several Dicotyledones which have been considered analogous to Apocyneæ, Euphorbiaceæ, Cacteæ, Coniferæ, &c. No great stress need be laid at present upon the several proportions which species of these classes bear to each other; as it is probable that subsequent researches will considerably modify them. The great predominance and size of arborescent ferns and other tribes of Ductulosæ constitute the main feature of the formation.

Above the coal we arrive at the new red sandstone; in some of the formations subordinate to this series a few species of fossil plants occur. In the oolitic series they become more abundant, and some beds are remarkably characterized by the prevalence of the genus Zamia, together with some Coniferæ, Liliaceæ, and many ferns, the latter being very distinct from those in the former formations. In the green sandstone and chalk few species have been hitherto found, and these are almost all marine. Among the tertiary strata (or those above the chalk) the Dicotyledones begin to prevail to a far greater extent than they did before, and the plants are entirely different, including terrestrial, lacustrine, and marine species. Several fruits are referable to existing genera, as Acer, Juglans, Salix, Ulmus, Cocos, Pinus, &c.

It is remarkable that scarcely any species has been found in more than one distinct formation, and none have occurred in any two which are separated by a long epoch. Hence it appears to be a natural conclusion, that there have been successive destructions and creations of distinct species. Mons. Brongniart has grouped the several formations in which vegetable remains are found, under four great epochs, during each of which no very marked transitions occur in the general character of the vegetation; but between any two of these epochs, a striking and decided change takes place: even most of the genera are different, and none of the species are alike. These epochs include the periods during which the following strata were deposited: —

1. From the earliest secondary rocks to the uppermost beds of the coal-measures.

2. The new red sandstone series.

3. From the lowest beds of the oolitic series to the chalk inclusive.

4. The beds above the chalk.

Judging from analogy, from the characters and relative proportions of the species in different classes, the temperature of those parts in which the plants of the first period were growing must have been both hotter and moister than the climates in any part of the earth at present. It has been plausibly conjectured that the atmosphere was more charged with carbonic acid at those early periods of our planet's history, when gigantic species of cryptogamic plants formed the main feature of its vegetation. The abundance of reptiles, also, without any Mammalia during the same epoch, appears favourable to this supposition. Since the fossil plants, which have been found in the arctic regions, are analogous to those which now grow in tropical islands, it seems likely, that not only must they have enjoyed a higher temperature, but also a more equable diffusion of light than those regions now possess. Speculations of this description, imperfect as they confessedly are at present, may one day lead to the most important results, and may teach us many truths respecting the earliest conditions of our planet, which the science of astronomy could never have suggested. And surely no one ought to consider such inquiries too bold for our limited faculties, needless for our present, or dangerous for our future welfare. No naturalist, desirous of knowing the truth, can be so weak as to fancy that any search into the works of God, or any contemplation of the wonders of his creation, can interfere with the lessons he has taught us in his revealed and written word. The commentator who wishes us to pay attention to his interpretations of the sacred text, must not proceed upon the supposition that there has been any thing written in the Bible for our learning, which can possibly be at variance with the clear and undeniable conclusions deducible from other and independent sources. If the letter does not announce a particular fact *revealed* in the works of the creation, a true believer will immediately infer that the letter (though it have the authority of inspiration) was not intended to teach that fact. When the philologist has ably interpreted the letter, the aid of the natural historian may still be needed before the divine can safely pronounce upon the exact scope and meaning of the instruction which it was intended to convey.

INDEX AND GLOSSARY.

The language of the botanist comprises many words adopted, or rather compounded, from Greek and Latin, which are seldom applied in their strictly classical signification; and some English terms are also employed in a peculiar and technical sense. The derivation of many of these is here given, that the reader may be the better able to remember them; but further reference is made to the article and page, where the fullest explanation of their meaning occurs, in the body of the work.

A.

Abortion (115.), 118.
Absorption (160.), 176.
Acotyledones (*α*, not; *κοτυληδων*, a seed leaf), (36.), 35.
Adfluxion (167.), 182.
Adventitious buds (57.), 51.
Aerial-stem (45.), 43.
Æstivation (*æstiva*, summer quarters), (104.), 101.
Age of trees (240.), 243.
Air-cells (21.), 19.
Air-cells (174.), 188.
Akenium (*α*, not; *χαινω*, to open), (108. 6. *fig.* 117.), 109.
Albumen (*albumen*, the white of an egg), (34. 1.), 32.
Albumen, formation of (269.), 271.
Alburnum (*alburnum*, sap-wood), (50.), 45.
Alternate (82.), 75.
Amnios (269.), 271.
Amylaceous (*amylon*, wheaten food), like flour.
Anastomose, (*αναστομωσις*, passing of one vein into another).
Anatropous (*ανα*, over; *τρεπω*, to turn), (267.), 271.
Angulinerved (72.), 62.
Annular (*annulus*, a ring), ringed.
Anther (*ανθηρος*, flowery), (97. and 98. *fig.* 98.), 96.
Apex (*apex*, the summit, pl. apices).
Apocarpous (*απο*, apart; *καρπος*, fruit), where the carpels are not united into a compound pistil, 103.
Arillus (109.), 111.
Articulation (69), 60.
Ascent of sap (163.), 178.
Assimilation (223.), 227.
Atmosphere, influence of (307.), 300.
Awn (96.), 96.
Axil (*axilla*, the arm-pit). The angle at which a leaf or branch unites with the stem.
Axis, imaginary line, drawn longitudinally through the middle of an organ.

B.

Bell-shaped, or campanulate (95. 1. *fig.* 92. *a*), 94.
Berry (108. 10. *fig.* 120.), 109.
Biennial, lasting two years.
Bladders (42.), 41.
Botanical geography (302.), 294.
Botanical habitations (309.), 302.
Botanical regions (312.), 304.
Botanical stations (308.), 301.
Bractea (*bractea*, a thin leaf of metal), (91.), 89.
Branches (59.), 51.
Budding (228. 3.), 233.
Buds (57.), 50.
Buds, on leaves (293. *fig.* 168.), 286.
Buds and embryos, connection of, (291.), 285.
Bulb (65.), 57.

C.

Caloric, development of (254.), 258.
Calycifloræ (102.), 101.
Calyx (*calyx*, the cup of a flower), (92. and 94.), 91.
Cambium (34. 2.), 32.
Camphor (208.), 218.
Campulitropous (καμπυλος, curved; τρεπω, to turn), (267.), 270.
Capitulum (*capitulum*, a little head), 90. *fig.* 87.), 89.
Capsule (*capsula*, a chest), (108. 8.), 109.
Cariopsis (καρη, the head; οψις, form), (108. 5.), 108.
Carpels (καρπος, fruit), (92.), 91. (100.), 98.
Catkin, (89. *fig.* 82.), 86.
Caudex (*caudex*, a stem), (39.), 38.
Caudex (84.), 77.
Caulinar (*caulis*, a stem), attached to the stem.
Cellulares (36. 2.), 36.
Cellular tissue (16.), 14.
Centrifugal inflorescence (88.), 84.
Centripetal inflorescence (89.), 86.
Chalaze (χαλαζα, tubercle in the skin), (266.), 270.
Chara (194. *fig.* 158.), 207.
Character (132.), 138.
Chromatometer (χρωμα, colour; μετρον, measure), (186.), 200.
Ciliæ (*cilium*, hair of the eyelids), fringes of hair or bristles, 167.
Circinate (*circinatus*, rounded), (*fig.* 72. *g*), 74.
Circulation (195.), 208.
Classes (33.), 30.
Closters (16.), (κλωστηρ, a spindle), elongated vesicles of the cellular tissue, 15.
Cluster. *See* Raceme.
Cohorts (131.), 137.
Colour (181.), 194.
Colour of fruit (274.), 275.
Complex organs (32.), 29.
Compound organs (28.), 24.
Cone (91. *fig.* 137.), 89.
Conduplicate (104.), 102.
Coniferous, bearing cones, as the fir tribes.
Connate (83. *fig.* 73. *a*), 75.
Connective (*connecto*, to join together), (98.), 97.
Conservative organs (10.), 10.
Contorted (*contortus*, twisted), (104.), 102.
Cormus (66.), 58.
Corolla (*corolla*, a little crown), (92. and 95.), 91.
Corollifloræ (102.), 101.
Corymb (κορυμβος, a summit, or a branch), (90. *fig.* 85.), 87.
Cotyledons (κοτυληδων, a hollow vessel); used in botany to signify the seed-leaves (34. 1.), 31.
Cow-tree (203. 3.), 216.
Crenate, cut into rounded teeth.
Cryptogamic (κρυπτος, concealed; γαμος, marriage), (36. 1.), 35.
Culms (*culmus*, a stem), the stem of grasses (96.), 96.
Curvinerved (73.), 66.
Cuticle (*cuticula*, the outermost skin), (29.), 25.
Cuticular, belonging to the skin or cuticle.
Cyma (*cyma*, a branch or sprout), (61.), 53.
Cyme (88.), 84.

D.

Deciduous (*deciduus*, liable to fall), opposed to persistent.
Decurrent (*decurro*, to run down), (83. *fig.* 74.), 76.
Degeneration (116.), 118.
Dehiscence (*dehiscens*, gaping), (107.), 105.
Depressed (*depressus*, pressed down), where the transverse section of an organ is larger than the longitudinal.
Descent of sap (190.), 204.
Deomodium gyrans (149. 2. *fig.* 150.), 166.
Development (230.), 234.
Diadelphous (δις, twice; αδελφος, a brother), (97.), 91.
Dichotomous (διχοτομος, divided in two), (88. *fig.* 80. *a*), 84.
Dicotyledones (δις, twice; κοτυληδων, a seed-leaf), (34.), 31.
Diffusion of proper juice (189.), 203.
Dionæa muscipula (149. 4. *fig.* 151.), 167.
Disk (101.), 99.
Dissemination (275.), 276.
Dissemination, modes of (279.), 278.
Dissepiment (*dissepio*, to separate), (106.), 104.
Divergent, separating asunder.
Divided. *See* Incised.
Divisions (131.), 137.
Drupe (*drupæ*, unripe olives), (108.) 3.), 108.
Drupel (108. 3.), 108.
Ducts (*ductus*, a pipe for water), (24.), 22.
Ductulosæ (36. 2.), 36.
Duramen (*duramen*, a hardening), (50.), 44.
Duration (235.), 238.

E.

Earths (220.), 224.
Eductulosæ (36. 2.), 36.
Elasticity of tissue (142.), 158.
Electricity (156.), 172.
Elementary textures (13.), 13.
Embryo (εμβρυον, the fœtus), (34. 1.), 31.
Embryo (111.), 112.
Embryo, formation of (268.), 271.
Embryo, vitality of (290), 285.
Embryonic sack (266.), 269.
Endocarp (ενδον, within; καρπος, fruit), (106.), 105.
Endogenæ (ενδον, within; γεινομαι, to beget), (35.), 33.
Endosmometer (144. *fig.* 148.), 160.
Endosmose (ἔνδον, within; ωσμος, impulsion), (144.), 159.
Ephemeral flowers (250.), 255.
Epicarp (επι, upon; καρπος, fruit), (106.), 105.
Epidermis (επιδερμις, the skin), (29.), 25.
Epigynous (επι, upon; γυνη, a woman), (101.), 100.
Epirrheology (επιρροη, an influx), (298.), 290.
Equinoctial plants (250.), 258.
Equitant (*equito*, to ride), (*fig.* 72. *b*), 74
Erect (111. *fig.* 126. *b*), 113.
Etiolation (178.), 192.
Excitability (146.), 161.
Excretions (212.), 220.
Exfoliate, to scale off.
Exhalation (168.), 185.
Exogenæ (εξω, without; γεινομαι, to beget), (34.), 31.
Expansion, stimulants to (251.), 256.
Extraneous matters (219.), 224.

F.

Farinaceous (*farina*, meal), formed of meal-like powder.
Fasciculate (*fasciculus*, a bundle), in bundles, (*fig.* 30. *c*), 41.
Fecula (197.), 211.
Fertilization (255.), 259.
Fibre (13.), 13.
Fibrils (39.), 38.
Filament (97.), 96.
Filamentous (*filum*, a thread), threadlike.
Fixation of carbon (175.), 189.
Flavour (273.), 274.
Flocculent (*floccus*, a lock of wool), wool-like.
Floral whorls (92.), 90.
Flower-buds (85.), 79.
Flower-buds (245.), 250.
Flowering (246.), 251.
Foliaceous branches (76.), 69.
Follicle (*folliculus*, a little bag), (108. 1. *fig.* 114.), 107.
Foramen (*foramen*, a hole), (111.), 113.
Foramen, (266.), 269.
Fossil Botany (315.), 310.
Fovilla (262.), 266.
Fraxinella (213.), 221.
Frond (*frons*, a leaf), (84.), 77.
Fruit (105.), 102.
Fugacious (*fugax*, fleet), lasting for a very short time.
Functions of vegetation (152.), 170.
Fundamental organs (38.), 37.
Funicular chord (*funiculus*, a little rope), (109.), 111.
Funnel-shaped, or infundibuliform, (95. 1. *fig.* 92. *b*), 94.
Fusiform (*fusus*, a spindle), spindle-shaped (*fig.* 3. *c*), 15.

G.

Gamosepalous (γαμος, marriage; *sepalum*, a sepal), where the sepals are united together.
Gemmule (*gemma*, a young bud), (111.), 113.
Genus (33.), 30.
Germen (*germen*, a bud). *See* Ovarium (100.), 98.
Germination (283.), 282
Germination, stimulants to (284.), 283.
Glans (*glans*, an acorn), (108. 7. *fig.* 118.), 109.
Glue (215.), 221.
Glossology (γλωσσα, the tongue; λογος, a discourse). The department of Botany which contains an explanation of the technical terms used in the science (3.), 3.
Glumaceous (96. *fig.* 95.), having the character of a glume, 95.
Glume (*gluma*, a husk of corn), (96.), 95.
Gluten, a tenacious substance extracted from flour.
Gourd (108. 9. *fig.* 119.), 109.
Grafts (227.), 231.
Granulated, having the appearance of being composed of grains.
Granules of the pollen (99.), 98.
Granules (263.), 267.
Gravity, effects of (300.), 292.
Growth (224.), 227.
Gum (177.), 191.

H.

Habitations (302.), 295.
Habitations (309.), 302.
Hair (31. *fig.* 19.), 27.
Heart-wood (50.), 44.
Heat, action of (287.), 285.
Herbaceous, of a soft and succulent nature — opposed to the woody structure of trees.
Hilum (*hilum*, the black on a bean), (109.), 111.
Hilum (266.), 270.
Horary expansion (250.), 255.
Hybrids (*hybrida*, a mongrel), (295.), 287.
Hygroscopicity of tissue (143), 159.
Hypocarpogean (ὑπὸ, beneath; καρπος, fruit; γη, the earth), (280.), 279.
Hypogynous (ὑπὸ, beneath; γυνὴ, a woman), (101.) 100.

I. & J.

Incised (*incisus*, cut), (*fig.* 63. *b*), 67.
Indefinite inflorescence (89.), 85.
Indehiscent (*in*, not; *dehiscens*, cleaving open), where there is no natural line of suture.
Individuality (236, 237, 238.), 239.
Inferior (101.), 100.
Inflorescence (86.), 80.
Inflorescence, stimulants to (247.), 251.
Intercellular (17.), 17.
Internodium (56.), the space between two knots, 50.
Inverse embryo (111., *fig.* 126. *a*), 113.
Involute (*involutus*, folded in), (*fig.* 72. *d*), 74.
Irritability (148.), 163.
Joints (56.), 50.

K.

Kernel (109.), 111.
Knot, (56.), 50.

L.

Labiate (*labium*, a lip), (95. 2. *fig.* 93.), 94.
Lacunæ (*lacuna*, a hollow place), (21.), 19.
Lamina, a thin plate of any thing.
Latex (*latex*, juice), (195.), 209.
Leaflets (70.) The subdivisions of a compound leaf, 61.
Left-handed spiral (55. *fig.* 41. *a*), 49.
Legume (*legumen*, pulse), (108. 2. *fig.* 115.), 107.
Lenticellæ (43.), 42.
Lenticular, shaped like a lens.
Light (154.), 171.
Light, action of (288.), 285.
Light, effects of (301.), 293.
Light, influence of (304.), 298.
Lignine (*lignum*, wood), (200.), 214.
Limb of a leaf (69.), 60.
Lime (220.), 224.
Linear, equally straight throughout, the edges parallel to each other.
Linnæan system (137.), 145.
Lipped. *See* Labiate.
Lobe, the separate divisions of a leaf or other organ, between the indentations on its margin.
Loculicidal (*loculus*, a little pouch), (107. *fig.* 111. *b*), where the opening is in the middle of the cell, 105.
Longevity of trees (241.), 244.
Lomentaceous (108. 2. *fig.* 115. *d*), where an organ, as the seed vessel, or a leaf, is much contracted at intervals, 108.
Lunate (*luna*, moon), crescent-shaped.
Lymph (*lympha*, water), (163.), 179.

M.

Macerate, to decompose by the action of water.
Maturation (265.), 268.
Maturation (271.), 273.
Maturation, stimulants to (272.), 274
Medullary rays (34. 2.), 33.
Medullary rays (51.), 45.
Medullary sheath (34. 2.), 32.
Medullary sheath (49.), 44.
Membrane (13.), 13.
Meteoric plants (250.), 256.
Migration, obstacles to (310.), 303.
Milk (203.), 215.
Moisture, action of (285.), 284.
Moisture, influence of (305.), 298.
Molecules (5.), the smallest particles (simple or compound) of which simple minerals are composed, 6.
Monadelphous (μονος, alone; αδελφος, a brother), (97. *fig.* 97. *a*), 97.
Monocarpean (μονος, alone; καρπος, fruit), (235.), 238.
Monochlamydeæ (μονος, alone; χλαμυς, a coat), 101.

Monocotyledones (μονος, alone; κοτυληδων, a seed-leaf), (35.), 33.
Monocotyledonous stems (53.), 46.
Monophyllous (μονος, alone; φυλλον, a leaf).
Monosepalous (μονος, alone; *sepalum*, a sepal).
Monstrosity (85.), 79.
Morphology (μορφη, form; λογος, a discourse), (114.), 116.

N.

Nectary (103.).
Nectary, functions of (253.), 258.
Nervation (71.), 61.
Nerves (69.), 59.
Nodosities, knotted appearances.
Normal (*normalis*, right by the rule), (115.), 118.
Nosology (νοσος, a disease; λογος, a discourse), (298.), 291.
Nucleus (266.), 267.
Nut (108. 4. *fig.* 116.), 108.
Nutrition (159.), 175.

O.

Obvolute (*fig.* 72. *f*), 74.
Oil (206.), 218.
Opposite (82.), 75.
Order (33.), 30.
Organizable products (176.), 190.
Organized bodies (6.), 6.
Organs (8.), 9.
Organography (οργανον, an organ; γραφω, to write), (3.), the department of Botany which contains a description of the organs of plants, 3.
Orthotropous (ορθος, straight; τρεπω, to turn), (267.), 270.
Ovarium and Ovary (*ovum*, an egg), (100.), the part of the pistil containing the seeds, 98.
Ovate (*ovum*, egg), egg-shaped, (*fig.* 30. *a*).
Ovule (*ovum*, an egg), (100.), the young seed, 98.
Ovule, development of (270.), 272
Ovule, modifications of (267.), 270.
Ovule, origin of (266.), 268.
Oxygen (180.), 193.
Oxygen, action of (286.), 284.

P.

Palmate (*palma*, the hand), hand-shaped, (*fig.* 30. *b*, and *fig.* 58.)
Palminerved (72. b.), 64.
Panicle (90. *fig.* 84.), 87.
Papilionaceous (*papilio*, a butterfly), (95. 3. *fig.* 94.), 95.
Parasites (234.), 235.
Parenchyma (69.), 59.
Paries (*paries*, the wall of a house), (parietes, *pl.*).
Parietal, belonging to the paries—attached to the paries.
Partite (*partitus*, divided), (*fig.* 63. *c*), 67.
Patent, spreading open widely.
Pedalinerved (72. d.), 65.
Pedate (*pes*. pl. *pedes*, a foot), (*fig.* 60.), a shape somewhat like a foot, 65.
Pedicel (86.), 80
Peduncle (86.), 80.
Pellicle (*pellis*, the skin), a thin skin.
Peltate (*pelta*, a shield), (*fig.* 59.), 65.
Peltinerved (72. c.), 65.
Penninerved (*pennatus*, winged), (72. a.), 63.
Perennial, lasting many years.
Perfoliate (*per*, through; *folium*, a leaf), (83. *fig.* 73. *a*, *b*), 76.
Perianth (περι, around; ανθος, a flower), (92, 93.), 90.
Perianth, functions of (252.), 257.
Pericarp (περι, around; καρπος, fruit), (106.), 103.
Perigynous (περι, around; γυνη, a woman), (101.), 100.
Periodic influences (249.), 254.
Periodicity (151.), 169.
Perisperm (περι, around; σπέρμα, seed), (269.), 271.
Permanence of species (296.), 288.
Persistent, remaining when other parts fall off.
Personate (*persona*, a mask, (95. 2. *fig.* 131. *a*), 94.
Petals (πεταλον, a leaf), (92.), the subordinate parts of the corolla, 91.
Petiole (*petiolus*, the stalk of fruits), (69.), used in botany for the stalk of leaves, 60.
Phanerogamic (φανερος, evident; γαμος, marriage), (36. 1.), 35.
Phyllodium (φυλλον, a leaf; εἶδος, form), (75.), 68.
Phytography (φυτον, a plant; γραφω, to write), (3.), the department of Botany which contains a description of the entire plant, 3.
Pinnate (*pinnatus*, feathered, winged), (72. a.), 63.
Pinnatifid (72. a.), cut in a pinnate manner, 63.
Pistil (*pistillum*, a pestle), (92. 100.), 92.
Pitcher (80.), 73.
Pith (34. 2.), 32.

Pith (48.), 44.
Placenta (100.), 99.
Placenta (105.), 103.
Plumule (*plumula*, a little feather), (34. 1.), 31.
Plumule (111.), 113.
Pollen (*pollen*, fine flower), (97. 99. *fig*. 99.), 96.
Pollen, dispersion of (258.), 262.
Pollen, formation of (261.) 265.
Pollen tube (262.), 266.
Polyadelphous (πολυς, many; αδελφος, a brother), (97.), 97.
Polycarpean (πολυς, many; καρπος, fruit), (235.), 238.
Polygonal (πολυς, many; γονη, an angle), having many angles and sides,
Pomum (108. 11. *fig*. 106. 121.), 110.
Preservation of seed (287.), 279.
Prickle (62.), 53.
Primary groups (33.), 29.
Primine (266.), 269.
Progression of sap (191), 205.
Proliferous (*proles*, the young; *fero*, to bear), (292.), 286.
Propagation (243.), 248.
Proper juice (202.), 215.
Propulsion (166.), 181.
Pruning (225.), 229.
Pseudospermic (ψευδος, a falsehood; σπερμα, seed), (276.), 277.
Pubescence (*pubescens*, downy), (31.), 27.
Pyxidium (πυξιδιον, a little box), (107. *fig*. 112.), 105.

R.

Race (131.), 137.
Raceme (*racemus*, a bunch), (89. *fig*. 81. *a*), 85.
Rachis (ραχις, spine of the back), (96.), 96.
Radical, proceeding from the summit of the root.
Radical excretions (217.), 222.
Radicle (*radicula*, a little root), (34. 1.), 31.
Radicle (111.), 113.
Raphe (ῥαφή, a joint or suture), (266.), 270.
Raphides (ραφις, a needle), (20.), 19.
Receptacles (21.), 19.
Receptacle to the flower (86.), 80.
Regions, botanical (312.), 304.
Reproduction (244.), 249.
Reproduction, certainty of (260.), 264.
Reproductive organs (11.), 10.
Resin (205.), 218.
Respiration (172.), 186.
Revolute (*revolutus*, turned back), (*fig*. 72. *e*), 74.
Rhizoma (ριζωμα, a root), (44.), 42.
Rhizoma (63.), 54.
Rhizoma (84.), 77.
Rhomboidal dodecahedron (*fig*. 5. *b*), a regular geometric figure, whose sides are twelve similar and equal rhombs, or plane four-sided figures, having their sides equal, but their angles not right angles, 16.
Rice-paper (50. *fig*. 36.), 45.
Right-handed spiral (55. *fig*. 41. *b*), 49.
Root (39.), 38.
Roots, direction of (299.), 291.
Rotate (*rota*, a wheel), wheel-shaped, (95. 1. *fig*. 92. *d*), 94.
Rotation (193.), 206.
Rotation of crops (218.), 223.
Runners (62.), 54.

S.

Salts (221.), 225.
Salver-shaped (or hypocrateriform), (95. 1. *fig* 92. *c*), 94.
Samara (108. 12. *fig*. 122.), 110.
Sarcocarp (σαρξ, flesh; καρπος, fruit), (106.), 105.
Scar (69.), 60.
Scent (210.), 219.
Scorpioidal (σκορπιος, a scorpion; ειδος, form), (88. *fig*. 80. *b*), 85.
Secretion (196.). 211.
Sections (131.), 137.
Secundine (266.), 269.
Seed (109.), 110.
Seed-cover (34. 1.), 31.
Sensibility (150.), 168.
Sensitive plant (149. 1. *fig*. 149.), 164.
Sepals (92.), the subordinate parts of the calyx, 91.
Septicidal (*septum*, a hedge or fence), opening along the divisions between the cells (107. *fig*. 111. *a*), 105.
Serrature (*serra*, a saw), having the edge jagged or toothed like a saw.
Sessile (*sessilis*, dwarfish), without a stalk.
Sexes (257.), 260.
Shoots (58.), 51.
Silica (220.), an earth; the basis of flints, quartz, &c., 225.
Siliqua (*siliqua*, a husk or pod), (108. 13. *fig*. 123.), 110.
Silver grain (51.), 45.
Simple mineral (5.), 6.

Sinus (*sinus*, a bay), the indentations on t' dge of a leaf.
Sleep (155.), 1.
Snag (225.), 229.
Soil, action of (289.), 285.
Soils, influence of (306.), 299.
Spadix (89. *fig*. 88. *b*), 86.
Spathe (σπαθη, a ladle), (91. *fig*. 88.), 90.
Species (33.), 29.
Spermoderm (σπερμα, seed; δερμα, skin), (109.), 111.
Spiculæ (*spiculum*, a dart), small thread-like and sharp-pointed bristles.
Spike (*spica*, an ear of corn), (89.), 85.
Spikelet, a little spike (89. *fig*. 95. *c*), 86.
Spine (*spina*, a thorn), (78.), 71.
Spiral-vessels (23.), 38.
Spongioles (*spongia*, a sponge), (39.), 38.
Sporules (σπορα, a seed), (36. 1.), the reproductive organs of the cryptogamic tribes, analogous to the seeds of flowering plants, 35.
Spur, the prolongation backwards of a sepal, petal, &c.,
Stamen (*stamen*, the chive of the flower), (92. 97.), 91.
Stations, botanical (302.), 294.
Stations, botanical (308.), 301.
Stellate (*stella*, a star), star-shaped, (*fig*. 21. *a*).
Stem (44.), 42.
Stems, direction of (299.), 291.
Stigma (100.), 98.
Stigma, action of (264.), 267.
Stings (31. *fig*. 20. *a*), 28.
Stings (214.), 221.
Stipes (*stipes*, trunk of a tree), (84.), 77.
Stipules (*stipula*, husk round straw), (77.), 70.
Stomata (στομα, the mouth), (30.), 26.
Stock (227.), 238.
Striated, marked with stripes.
Style (στυλος, a style), (100.), 98.
Suckers (62.), 54.
Sugar (199.), 213.
Superior (101.), 100.
Suture (*sutura*, a seam), where a division takes place naturally in the fruit.
Syncarpous (συν, together; καρπος, fruit), (*fig*. 106.), 103.
Syngenesious (συν, together; γενεσις, generation), (138.), 149.

T.

Tap (39.), 38.
Taste (210.), 219.
Taxonomy (ταξις, order; νομος, a law), (130.), the same as systematic Botany. — The Department of the science in which plants are arranged and classified, 135.
Tegmen (*tegmen*, a covering), (266.), 269.
Temperature (157.), 172.
Temperature, effects of (303.), 295.
Tendril (79.), 71.
Terminal inflorescence (88.), 83.
Testa (*testa*, an earthen pot), (266.), 269.
Thalamifloræ (θαλαμος, a bed-chamber), (102.), 101.
Thallus (84.), 78.
Theca (θηκη, a sheath or case), (113.), 115.
Thecaphore (θηκη, a case; φερω, to bear), (100.), 99.
Thorns (62.), 53.
Toothed (*fig*. 63. *a*), 67.
Torus (*torus*, a bed), (92.), 90.
Tracheæ (23.), 21.
Transport, means of (311.), 304.
Transverse, embryo (111. *fig*. 126. *c*), 113.
Tribes (131.), 137.
Tuber (*tuber*, an excrescence), (64.), 56.
Turio (*turio*, a young branch), (58.), 51.

V.

Valve, a part which becomes detached by means of a natural rupture along a line of suture, as in seed-vessels.
Valvular (104.), 102.
Variation (131.), 137.
Varieties (33.), 30.
Varieties (131.), 137.
Varieties, origin of (297.), 290.
Vasa propria (21.), 19.
Vasculares (36. 2.), 36.
Vascular tissue (22.), 20.
Veins (69.), 59.
Venation (71.), 61.
Vernation (*vernus*, belonging to spring-time), (81.), 74.
Verticillate (*verticulum*, a whirl for a spindle), (82.), 75.
Vesicles (*vesicula*, a little bladder), (16.), 14.
Viscous (*viscus*, glue), clammy and glutinous.
Vital vessels (27.), 24.

U.

Umbel (90. *fig.* 86.), 87.
Umbellate, in the form of an umbel.
Umbilical chord. *See* Funicular.
Under-shrub (45.), 43.
Unorganized bodies (5.), the objects of the mineral kingdom, 5.

W.

Wax (216.), 222.
Winged (83. *fig.* 74.), 76.
Wood (50.), 44.
Woody fibres (25.), 23.

THE END.

Printed in the United States
By Bookmasters